Separatdruck aus «technica» Nr. 4 und 5 (1965),
Seite 247 und 355

Nachdruck verboten
Alle Rechte, insbesondere das der Übersetzung in fremde
Sprachen und der Reproduktion auf photostatischem Wege
oder durch Mikrofilm, vorbehalten

ISBN 978-3-0348-4051-4   ISBN 978-3-0348-4124-5 (eBook)
DOI 10.1007/978-3-0348-4124-5

©
Springer Basel AG 1965
Ursprünglich erschienen bei Birkhäuser Verlag, Basel 1965.

Reprinted and translated from "technica", Nos. 4 and 5 (1965),
pp. 247 and 355

All rights reserved, including the right of translation and
of reproduction by photostatic means or on microfilm.

©
Springer Basel AG 1965
Originally published by Birkhäuser Verlag, Basel in 1965.

Zwanzig Jahre ARALDIT-Funktionserfindung

The invention of ARALDITE: 1944–1964

Die Bindefunktion der Äthoxylin-(Epoxy-)Harze
von Dr. Eduard Preiswerk, Basel

The recognition of the bonding properties of epoxy (ethoxyline) resins
by Eduard Preiswerk, Ph. D., Basle

«technica»-Reihe, Nr. 9
Springer Basel AG
1965

Inhalt	Contents
Zusammenfassung	Summary
1. Die neuerkannte Bindefunktion der Äthoxylin-(Epoxy-)Harze	1. The recognition of the "bonding function" of epoxy resins
2. Der Stand der Technik zum Zeitpunkt der ARALDIT-Funktionserfindung (Bindefunktion) im Jahre 1944	2. The state of adhesives technology in 1944
3. Die Entstehung der Äthoxylin-(Epoxy-)Harze	3. The origin of epoxy resins
4. Die Äthoxylin-(Epoxy-)Harze als zahnärztliches Prothesenmaterial	4. Epoxy resins in prosthetic dentistry
5. Erste Überprüfungen von de Treys SELITROL-Harzen durch die Technik	5. First attempts to investigate de Trey's SELITROL resin
6. Das ausserordentlich gute Haften an Glas, Porzellan und Metallen und das Nichterkennen der dann erfinderisch offenbarten Bindefunktion	6. Bonding function unrecognised despite outstanding adhesion to metals, ceramics and glass
7. Das Entstehen der ARALDIT-Funktionserfindung	7. The "invention of ARALDITE" (the discovery of the bonding function)
8. Die weitere Bewährung der Äthoxylin-(Epoxy-)Harze	8. Further experience with epoxy resins
9. Namen, Bezeichnungen und Begriffe im Zeichen des technischen Fortschrittes	9. Terminology
10. Die Bedeutung der Bindefunktion der Äthoxylin-(Epoxy-)Harze in der Praxis der Technik	10. The importance of the bonding function in technological practice
11. Bestrebungen der internationalen wissenschaftlichen Forschung zur Bestimmung der Bindefunktion der Äthoxylin-(Epoxy-)Harze	11. The bonding function of the epoxies scientifically investigated

Zusammenfassung

Anlässlich des 20jährigen Bestehens der sog. ARALDIT-Funktionserfindung (1944) werden die wesentlichen Tatsachen aus der Geschichte dieser Werkstoffentwicklung bekanntgegeben.

Gehärtete Kunstharze vom Typus der Äthoxylin-(Epoxy-)Harze mit all den diese Harzklasse charakterisierenden Eigenschaften waren erstmals im Jahre 1938 von der Firma Gebr. de Trey AG in Zürich (Erfinder: P. Castan) entwickelt und zum Patente angemeldet worden. Im Jahre 1940 wurde das schweizerische Patent bekanntgemacht. In der gleichen Zeitperiode war auch ein Harz/Härter-Gemisch als allgemein käufliches Produkt zur Herstellung von Zahnersatzstücken wie Gaumenplatten, Gebissen usw. unter der Wortmarke SELITROL in den Handel gekommen. In den USA arbeitete die Lackfirma Devoe & Raynolds, welche dann mit der Firma Shell in enger Zusammenarbeit stand, seit 1941 an der Entwicklung des Äthoxylin-(Epoxy-)Harzes. Diese Arbeit konzentrierte sich indessen praktisch allein auf die Schaffung von Lackharzen. Seit Ende November 1942 prüfte die CIBA Aktiengesellschaft, Basel, dessen Verwendung für die Technik. Die Erkenntnis des aussergewöhnlichen Metallbindevermögens und dessen grundlegenden applikatorischen Funktion wurde Sonntag, den 26. November 1944 durch E. Preiswerk einerseits und durch A. Gams, CIBA, anderseits eröffnet. Sie sicherte in der Folge dieser Harzklasse unter der Wortmarke ARALDIT hervorragende Anwendungsmöglichkeiten in allen Sparten einer weltweiten Technik. Die neuerkannte Funktion erlaubte im Vereine mit den übrigen bereits festgestellten mechanischen, elektrischen und chemischen Standardeigenschaften erstmals das Verbinden und Vereinigen der verschiedensten Werkstoffe, insbesondere von Metallen und Glas, mittels Kunstharzen in einem mechanisch und elektrisch höchstwertigen Verbande, ohne dass Massnahmen wie die Anwendung eines spezif. Pressdruckes während des Verbindungsprozesses erforderlich und ohne dass der Dicke der Verbindungsfuge Grenzen gesetzt waren. Es wurde damit eine Entwicklung eingeleitet und entscheidend gefördert, die heute als «Technik der Verbundkörper» («composites») von andauernd wachsender Bedeutung ist. Die internationale, massgebende technische Forschung strebt danach, diese Bindekraft nicht nur unter idealisierten Verhältnissen und an eigens dazu hergestellten Proben zu messen, wie sie die klassischen Scherfestigkeitsprüfungen an überlappend verklebten Blechstreifen darstellen, sondern dieselbe am fertigen Applikationsobjekte selbst zu untersuchen.

Summary

To mark the twentieth anniversary of the "invention of ARALDITE," an account is given of the principal landmarks in the history of the development of this material.

Cured resins of the ethoxyline (epoxy) type, possessing all the various properties that characterise this class of resins, were first developed in 1938 by the firm of Gebr. de Trey AG of Zurich (inventor: P. Castan), when the relative patent application was filed. The Swiss patent was subsequently published in 1940. During the same period a resin/curing agent mixture was made generally available on the market under the trade name of SELITROL for the preparation of dentures. In the United States the paint and varnish manufacturing firm of Devoe & Raynolds, which at that time was working in close collaboration with the Shell Chemical Corporation, had also been engaged since 1941 in the development of ethoxyline (epoxy) resins. This work, however, was confined almost exclusively to the development of resins for surface coatings. In November 1942 CIBA Limited of Basle began to investigate the technical applicability of the resin. Its outstanding metal-bonding properties, or function, of fundamental significance for technical applications, were recognised on Sunday 26 November 1944 by E. Preiswerk on the one hand and by A. Gams, CIBA, on the other, and as a result this class of resins, to which the trade name ARALDITE was given, was assured of an exceptional and world-wide application potential in all branches of technology. Together with the already known standard mechanical, electrical and chemical properties of the resin, its newly discovered bonding function meant that for the first time it became possible to unite and combine a wide range of materials, in particular metals and glass, so as to produce units of optimum mechanical and electrical characteristics without having to resort to such measures as the application of a specific moulding pressure during the bonding process and without having to impose any limitations on the thickness of the actual joint. In this manner a field of development, which is today referred to as "composites" technology and which is of constantly growing importance, was given its original impetus and significantly furthered. Today leading research institutions throughout the world are engaged in investigating the properties of this material not only by measuring its bonding strength under idealised conditions with the aid of test specimens specially prepared for the purpose, as exemplified by the classical shear strength tests with bonded overlapping metal strips, but also by examining its behaviour in finished articles for specific applications.

Epoxy resins: definition

"Ethoxyline resins (the name is derived from ethylene oxide), also known as epoxide or epoxy resins, contain on the average more than one ethylene oxide group per molecule. The ethylene oxide group that typifies this class of resins possesses unusual reactivity. Once the three-membered ring has been opened, the group readily enters into polyaddition reactions with compounds containing an active hydrogen atom, such as amines, acids, phenols, mercaptans, and alcohols. When compounds containing more than one epoxide group are reacted with compounds containing several active hydrogen atoms the result is a highly cross-linked polymer. The transition to the cured state takes place exclusively via polyaddition reactions: no low-molecular-weight substances whatever are split off. Shrinkage on curing is remarkably low, and the resulting product possesses excellent cohesion and outstanding adhesion to substrates of all kinds. Epoxy resins possess a *combination*[1] of properties that make them extremely valuable in plastics technology, and this explains why, despite their relatively high price, they have undergone such rapid development and are currently used in so many applications in virtually all sectors of industry."
(W. Fisch, CIBA, Basle, in Vol. II of "Chemie und Technologie der Kunststoffe", Houwink-Staverman, 4th Ed., Leipzig 1963).

The above brief definition and description of epoxy resins speaks of a *combination*: a combination of properties in the literal sense, but also, in our present context, the result of a process of "bisociation", or the rare "moment of truth" when two previously unconnected frames of reference connect to create a new idea[2]. It was this correlation of facts and ideas that provided the stimulus for and at the same time represented the essence of the discovery on 26 November 1944 of the outstanding adhesive properties of epoxy resins and was subsequently acknowledged as the "invention of ARALDITE". The existence of ethoxyline resins had already been discovered in 1938 and made public in 1940, and the resin had indeed already been commercially available from the dental supplies firm of Gebrüder de Trey AG in Zurich (where they had been invented by Dr. P. Castan), but it was the recognition of the *combination* of properties referred to above that aroused interest in them throughout the world, provided the impetus for their further development, and was ultimately responsible for their application in virtually every branch of modern industrial technology. The twentieth anniversary of the "invention of ARALDITE", which took place in Switzerland, or more precisely in Basle [1][3], has just passed, and it is therefore not inappropriate to mark the event by giving a brief account of how the invention came about [2].

1. The recognition of the "bonding function"[4] of epoxy resins

The first specimens of metal parts bonded with ethoxyline, or epoxy, resins were inspected by scientists and technologists from all over the world at the stand of CIBA Limited at the Swiss Industries Fair in May 1946, the first such fair to be held after the war [3]. Sheets of light alloy bonded together with epoxy resin were seen being subjected to repeated flexural fatigue strength tests in a machine designed by the Versuchsanstalt für Luftschiffahrt (now renamed the Deutsche Forschungsanstalt für Luft- und Raumfahrt e.V. – the German Aerospace Research Establishment), and visitors were able to observe the performance of the sample while the test was actually in progress [4]. A large coloured reproduction of a Wöhler stress-cycle diagram [5] was also provided to demonstrate that the metal-to-metal bond obtained in this manner was of greater strength even than that obtainable by riveting (Fig. 1), and this remarkable result could be observed as the test was conducted in full view of the visitors. Other exhibits on the CIBA stand demonstrated how this new synthetic resin-based bonding agent could be introduced into the space where it was to perform its bonding function by simple casting,

---
[1] Author's italics (cf. also "Der Grosse Brockhaus" Encyclopaedia, p. 497, Vol. 6).
[2] See "The Act of Creation" by Arthur Koestler, Hutchinson, London, 1964 (p. 120).
[3] See page 21 for notes and references.
[4] The term "invention of a function" ("Funktionserfindung") is today a precisely defined and well established concept in the legal terminology of the German-speaking countries. Since there is no exact equivalent in the English language, the literal translation of the German term is retained in the present article for the sake of convenience.

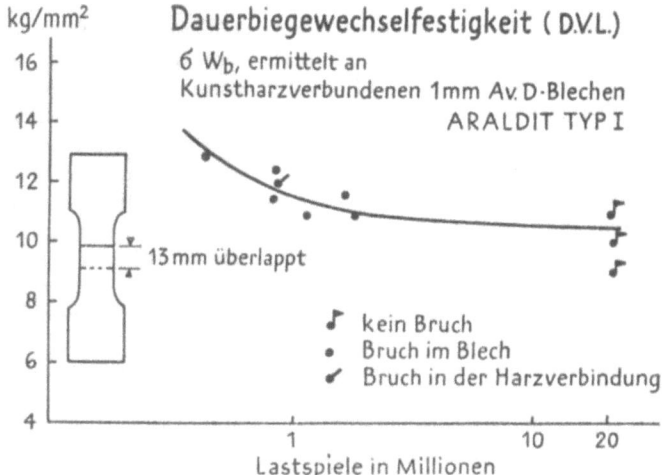

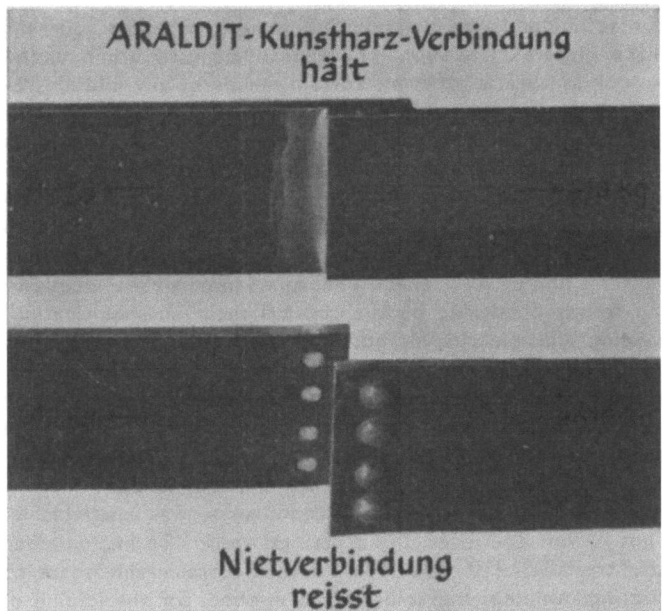

Bild 1. Mai 1946: Zum ersten Male wird der internationalen Fachwelt die neue, höchstwertige Bindefunktion der Äthoxylin-(Epoxy-)Harze bei statischer und dynamischer Beanspruchung bekanntgemacht. Das erste ARALDIT-Verkaufsprodukt wird den Interessenten angeboten.
(Aus: Aufsatz Preiswerk und von Zeerleder, «Schweizer Archiv für Angewandte Wissenschaft und Technik», Heft 4 (April), 12. Jahrgang, 1946, Abb. 3 und 7)

Fig. 1. May 1946: the newly discovered bonding function of ethoxyline (or epoxy) resins under conditions of static and dynamic (alternating fatigue) loading was announced to scientists and technologists for the first time, and the first commercial grade of ARALDITE was placed on the market.
(From an article by Preiswerk and von Zeerleder in "Schweizer Archiv für Angewandte Wissenschaft und Technik", No. 4 [April], Vol. 12, 1946, Figs. 3 and 7.)

Löten, Schweissen, Kitten, Compoundieren u.a.m. ergänzen, kombinieren oder ersetzen. Vor allem schuf sie die völlig neue Erkenntnis, dass, wo immer auch ein Äthoxylin-(Epoxy-)Harz zwischen zwei Werkstoffflächen gleich welcher Gestaltung und Distanz voneinander – vornehmlich von solchen nichtporöser Materialien wie Metall, Glas, Keramik usw. – zur Härtung gebracht wurde, eine höchstwertige, kraftschlüssige Verbindung entstand [6]. Die funktionelle Güte einer solchen Verbindung war an Hand von überlappend verklebten Metallblechsteifen unter idealisierten, übersichtlichen und reproduzierbaren Verhältnissen anlässlich jener ersten öffentlichen Schaustellung im Mai 1946 eindrücklich mit Messwerten unter statischen und dynamischen Bedingungen belegt worden. Bis auf den heutigen Tag blieb diese Prüfart die naheliegende Methode der Wahl, um die Güte der Bindefunktion irgendwelcher Äthoxylin-(Epoxy-)Harze festzustellen. Millionenfach kündeten in den seither verflossenen

impregnating, or similar processes. The resin, liquid in its original form, was cured without requiring the application of pressure. The manner in which these resins were used was reminiscent of metal soldering techniques.

The critical observer could immediately perceive that epoxy resins represented a totally new, original material of virtually limitless potential. Here was the long sought-after plastic that could be used for the bonding of all types of substances, but especially of glass and metals, to produce a "composite" material possessing outstanding mechanical and electrical properties. The new method was capable of being combined with, or of replacing, the more familiar techniques of riveting, soldering, welding, cementing, "compounding", and many others. In particular, the CIBA exhibit revealed the totally new concept of a bonding agent that could be used for joining any two surfaces, whatever their shape or the distance between them, to provide, after curing, a high-quality composite of excellent mechanical strength [6]. This had obvious implications for such non-porous materials as metals, glass, and ceramics. The functional quality of composites of this type was impressively confirmed with the aid of precise static and dynamic data obtained by testing overlapping epoxy-bonded metal strips ("lap joints") under idealised and reproducible conditions during the first public demonstration in May 1946. This classical test procedure has until today retained its place as the method of choice for investigating the bonding strength of epoxy resins of all types: in the intervening years the bonding function of these resins has been confirmed a million times over by the recording dials of testing machines.

Today, in virtually every field of technology, the engineer is turning to an ever-increasing extent to epoxies where the basic requirement is the *combination* of properties mentioned at the beginning of this article, i.e. the bonding function that was invented in 1944. The quality, and in particular the reliability of the bond that can be achieved with these resins has been undisputed from the moment when the epoxies were first presented to the public in May 1946 [84].

At the opening of the German Plastics Institute at Darmstadt on 19 June 1957, Professor Richard Vieweg, the President of the Federal Institute of Physics and Technology in Brunswick and Berlin, delivered an inaugural address entitled "Plastics: Past, Present, and Future" to an audience which consisted of leading German scientists and technologists. He said [7]:

"There is little point in reminding you of the many familiar applications of plastics in the engineering, textile, and paper industries. I need hardly mention that today we find plastics in everything from cord pulleys to gearwheels, from slideways to brake and clutch linings, from pipelines to mock-ups of apparatus. Medicine and surgery are two further fields that have benefited enormously from plastics. I should like to make special reference to only two of the forms in which plastics are available to us. First of all plastic foams, which have recently acquired such great importance....The two directions that this development has followed, namely lightweight construction and sound insulation, would be much less promising if we had not simultaneously made progress in another field: that of adhesives. For now we are able to bond metal to metal, to plastics, or to virtually any kind of surface, and obtain a lasting and reliable bond every time. Adhesives have even acquired major importance in construction engineering: plastics have become indispensable to metals technology, but at the same time we must not forget that plastics cannot exist without metals...."

Earlier, in 1951, Professor Vieweg had pointed out [8]:

"Plastics of the type sold as metal adhesives or under similar designations have now been perfected to the stage where they can be used for joining metal to metal or to almost any other kind of surface to produce a bond of a strength that until only recently would have been considered by many people as a physical impossibility" [9].

2. The state of adhesives technology in 1944

Franz M. Feldhaus, the well-known authority on the history of technology, mentions in his book "Die Technik der Antike und des Mittelalters" ("The Technology of the Ancients and of the Middle Ages", Akad. Verlagsgesellschaft, Potsdam 1931) that in about the year 530 B.C. Theodoros of Samos not only developed

Jahren – wo immer diese Harze geprüft werden – die Messuhren der Prüfmaschinen von dieser mittels Scherfestigkeiten bestimmten Funktion.

In der vielseitigen Praxis der modernen Technik zog und zieht der verantwortliche Konstrukteur in immer steigendem Masse das Äthoxylin-(Epoxy-)Harz einem Konkurrenzharz in der Regel dann vor, wenn ihm die einleitend genannte Kombination von Eigenschaften, d. h. die erfinderisch erkannte Bindefunktion wesentlich ist. Von Anfang an war ihm die Güte der Verbindung, insbesondere deren Zuverlässigkeit (im englischen Sprachgebrauch: reliability) in überzeugender Weise aufgezeigt worden [84].

Anlässlich der Einweihung des Deutschen Kunststoffinstitutes in Darmstadt am 19. 6. 57 wandte sich der Präsident der Physikalisch-Technischen Bundesanstalt in Braunschweig und Berlin, Prof. Dr. Richard Vieweg, in seiner festlichen Ansprache unter dem Titel «Kunststoffe, gestern, heute und in Zukunft» mit folgenden Worten an die anwesenden Spitzen der deutschen Technik [7]:

...«Die vielen bekannten Anwendungen von Kunststoffen im allgemeinen Maschinenbau, in der Textilindustrie, der Papierindustrie, wollen wir nicht einmal streifen. Es wäre trivial, heute darauf einzugehen, dass wir Kunststoffe von der Schnurrolle bis zum Zahnrad, von der Gleitbahn bis zur Kupplung und Bremsbelag, von der Rohrleitung bis zum Apparatemodell finden. Und wer möchte die medizinische Technik vergessen? Wir wollen nur noch zwei spezifische Erscheinungsformen von Kunststoffen erwähnen. Das sind die Schäume, die neuerdings eine grosse Bedeutung erlangt haben. ... Die beiden Entwicklungslinien des Leichtbaues und der Entdröhnung wären kaum so aussichtsreich, wenn nicht gleichzeitig ein anderer Problemkreis grosse Förderung erfahren hätte, die Klebung. Weithin interessieren diese neuen Möglichkeiten Metalle untereinander und Metalle und Kunststoffe wechselseitig und mit fast beliebigen Partnern sicher und dauernd haltbar zu verbinden. Sogar als konstruktives Element fordert das Kleben heute Beachtung. Wenn aus der Metalltechnik die Kunststoffe nicht mehr wegzudenken sind, so soll auch nicht vergessen werden, dass umgekehrt ohne Metalle die Kunststoffe nicht sein können»...

Schon 1951 hatte Prof. Vieweg festgestellt [8]:

...«Eben haben die als Metallkleber oder unter ähnlichen Bezeichnungen bekannt gewordenen Kunststoffe Leistungen in der Bindekraft zwischen Metallen oder zwischen einem Metall und einem fast beliebigen Partner vollbracht, die bis vor kurzem noch von manchem für grundsätzlich unmöglich gehalten wurden»... [9].

2. Der Stand der Technik zum Zeitpunkt der ARALDIT-Funktionserfindung (Bindefunktion) im Jahre 1944

Der bekannte Historiker der Technik, Franz. M. Feldhaus, berichtet in seinem Werke «Die Technik der Antike und des Mittelalters» (Akad. Verlagsgesellschaft, Potsdam 1931), dass Theodoros von Samos um 530 v. Chr. nicht nur die Technik des Giessens verbessert, sondern auch eine neue Technik, die man als das Zusammenleimen von Metall an Metall übersetzen muss, erfunden haben soll. Auch seien im Jahre 35 v. Chr. in Rom Bleirohre für Wasserleitungen verkittet worden. Eine Art «oberflächliche Vergiessung mit Kitt», wie Feldhaus vormerkt. Solche Angaben zeigen, dass der handwerklich und technisch suchende Mensch schon frühzeitig versuchte, mittels der bei andern nichtmetallischen Werkstoffen üblichen Verbindungsmittel (Leimen, Kitten usw.) auch bei Metallen eine Lösung zu finden. Obwohl in der Folge ganz offensichtlich das Metallverbinden, insbesondere das kraftschlüssige, sich der Methoden des Verbindens mittels Materialien der eigenen Werkstoffgattung, wie Nieten, Schweissen, Hart- und Weichlöten bediente, wurde doch immer versucht, mittels nichtmetallischer Verbindungsmittel zu arbeiten. So war es auch natürlich, dass Baekeland zu Beginn des Jahrhunderts in seinem klassischen Phenolharzpatent für die Anwendung seines neuen Werkstoffes auch das Metallverbinden postulierte. Dass auch dieser Versuch in der Folge praktisch von keiner Bedeutung war, bewies die weitere Entwicklung. Ja, es zeigte sich, dass die Technik ein wohlbegründetes Vorurteil gegen das Verkleben/Verleimen von Metallen besass. Obwohl immer wieder versucht wurde, mittels organischer Substanzen Metalle (Glas, Keramik usw.) zu verbinden, und sich dieser Wunsch seit dem Aufkommen der

improvements in the art of casting metals but also invented a new technique that can only be translated as "glueing metal to metal". We also read that lead water-pipes were being cemented together in Rome in the year 35 B.C.: a sort of "surface casting with cement", as Feldhaus describes it. Historical records such as these confirm that even in ancient times the inventive mind of man was already at work trying to apply, to metals, adhesives such as glues and cements that were already in common use for bonding non-metallic materials. Although in the following centuries the joining of metals, particularly where high mechanical strength was required, could be done only by using metallic materials, i.e. by riveting, welding, soldering, and brazing, attempts were nevertheless constantly being made to discover some non-metallic substance that could be used for this purpose. It is thus understandable that Baekeland, when he published his classic patent for phenolic resins at the beginning of the century, suggested that his new material might also be used for joining metals. Further development, however, showed this new technique to be disappointing. Indeed, it became apparent that engineers themselves, not without some justification, tended to look askance at attempts to join metals by any method requiring the use of an adhesive. Although efforts were repeatedly made to bond metals, glass, ceramics, and similar materials with the aid of organic substances, and although interest in such a possibility was intensified through progress in synthetic organic chemistry, the results were unsatisfactory: from time to time a bonding agent was developed that was capable of rendering some limited service, but it remained impossible to duplicate the results obtainable by the classical techniques of riveting, welding, and soldering. Although virtually every synthetic resin developed was recommended as a bonding agent for metals, every chemist was only too aware that none of them was ever seriously considered as a "synthetic resin solder". The adhesive bonding of non-porous materials such as metals and glass with synthetic resins – analogous to the bonding of porous materials with the traditional adhesives – was unattainable. Thus A. von Zeerleder, in his classical work "Technologie der Leichtmetalle" (1st Ed., 1936), devotes considerable space to riveting, welding, and soldering as methods for the joining of metals but makes no mention whatever of adhesives [10]. In the summer of 1940 a detailed and critical analysis entitled "Adhesives and Cements for Metals" [11] was written, in which it was stated that while adhesives might play a minor and subordinate role they would never, in the eyes of the engineer, compare with the standard techniques of soldering, brazing, welding, and riveting; in the view of the author there was little or no prospect that a synthetic resin-based bonding agent would ever be developed capable of wholly or partially replacing such traditional methods [12], indeed it was difficult even to envisage the chemical and physical nature of a resin that might be suitable for the purpose [13]. Such authoritative opinions as these help us to understand why the problem was regarded as impossible of solution [8, 9]. World War II provided further stimulus to renewed research, especially in the Allied countries, with the aim of developing a method that could replace riveting. Products were developed – they were in fact modified phenolic resins (e.g. P.F./acetals, P.F./nitrile rubber, and P.F./nylon), though the fact was not revealed for security reasons – which, as far as could be gathered from the scanty and somewhat sensationalistic information that was eventually published [14], required the application of pressure and did indeed seem capable of replacing rivets under certain conditions, though the process could still not be compared with welding, soldering, and similar techniques. The new resins, while excellent and representing a real advance in certain specialised applications (e.g. aircraft construction: "Dove", "Comet", "C.V. Cutlass", and many others) [15], were nevertheless of relatively limited application potential. A genuinely original, universally usable synthetic bonding agent that would enable metals to be "soldered" together still remained to be invented. And it was evident that an invention of this kind was

synthetischen organischen Chemie besonders an deren Adresse richtete, so gelang ein solches Verkleben wohl hie und da mehr oder weniger behelfsmässig, aber es war in keiner Weise vergleichbar mit den klassischen Verbindungsmethoden wie Nieten, Schweissen oder Löten. Die organische Chemie wusste, dass bis anhin – trotzdem beinahe jedes neue Kunstharz auch als Metallbindemittel empfohlen wurde – keines dieser Harze als Metallbindemittel bzw. als «Kunstharzlot» zu ernsthaftem Gebrauche in Betracht gezogen wurde. Kunstharzverbinden von Metallen (Glas, Keramik usw.), im Sinne eines klassischen Verbindungsverfahrens, war unbekannt. In dem Lehrbuch «Technologie der Leichtmetalle» von A. von Zeerleder, 1. Auflage, 1936, fand z.B. neben den ausführlichen Beschreibungen der üblichen Verbindungsverfahren, wie Nieten, Schweissen und Löten, das Kleben von Metallen noch mit keinem Worte Erwähnung [10]. Im Sommer 1940 wurde in einem umfangreichen und die Sachlage kritisch analysierenden Aufsatze «Adhesives and Cements for Metals» festgestellt [11], dass nach der Auffassung des Ingenieurs Kitte und Klebstoffe gegenüber den traditionellen Methoden des Weich- und Hartlötens, des Schweissens und Nietens, nur eine sekundäre Rolle spielen können und dass eine Entwicklung von Kunstharzbindemitteln, welche diese Methoden ganz oder teilweise ersetzen würden, nicht vorauszusehen sei [12]. Es sei überhaupt schwierig irgendwie vorauszusagen, wie Harze, die eine solche Aufgabe erfüllen könnten, beschaffen sein müssten [13]. Solch kompetente Aussagen beleuchteten eine Situation, welche der Fachmann mit dem Begriffe «unmöglich» kennzeichnete [8, 9]. Wohl waren im Verlaufe des Krieges in den angelsächsischen Ländern einmal mehr Anstrengungen unternommen worden, um Kunstharze als Metallbindemittel zu prüfen. Der Hauptzweck war, die Nietung zu ersetzen. Solche Produkte – es wurde geheimgehalten, dass es plastifizierte Phenolharze waren – verlangten, soweit es dem wenig und eher sensationell Bekanntgemachten [14] entnommen werden konnte, Druckanwendung beim Verbindungsprozess und schienen auf diese Art in gewissen Fällen den Nietprozess, nicht aber das Schweissen, Löten oder ähnliche Vorgänge ersetzen zu können. Solchen Verfahren, so erfolgreich sie sich dann etwa für einen Einzelfall entwickeln mochten [15], blieb denn auch in der Folge ein nur sehr beschränkter Anwendungsraum übrig. Das neuartige, vielseitig verwendbare Kunstharzbindemittel mit «Kunstharzlotcharakter» war noch zu erfinden! Wie schwierig die Tätigung einer solchen Erfindung zu sein schien, geht daraus hervor, dass John Delmonte, der bekannte nordamerikanische Spezialist für Leime und Klebstoffe, noch im Dezember 1946 – zwei Jahre nach dem Erfindungsdatum vom 26. 11. 44 – in seinem Handbuche «The Technology of Adhesives» [16] bei der Beschreibung der angelsächsischen Metallbindemittel CYCLEWELD und REDUX mitteilte: «... Various others have been developed from time to time... It would seem that the goal for the ideal adhesive still lies ahead. Most of the strong metal to metal bonding types require... tooling costs of applying the correct heat and pressure for curing.» Delmonte kannte damals noch nicht «CIBA's ARALDITE metal bonding types», welche im Frühjahr 1946 in den Handel gekommen waren. Diese Produkte verlangten allerdings keine «tooling costs of applying the correct... pressure for curing». Sie stellten augenscheinlich Delmontes «goal for the ideal adhesive» dar! Und welche Ironie des Schicksals besteht darin, dass in den USA selbst die Lackfirma Devoe & Raynolds, welche dann mit der Firma SHELL in enger Zusammenarbeit stand, schon seit 1941 [17], an der Entwicklung des Äthoxylin-(Epoxy-)Harzes gearbeitet haben soll. Diese Arbeit konzentrierte sich aber auf die Schaffung von Äthoxylin-Lackharzen und führte damit von Anfang an zu jener beherrschenden Stellung von SHELL und Devoe & Raynolds auf dem Gebiete des Oberflächenschutzes. Eine andere Entwicklungsrichtung fand daneben vorerst keine nennenswerte Pflege.

Das Überraschende ist – nachträglich besehen –, dass dieses Äthoxylin-(Epoxy-)Harz bereits im Jahre 1938 in der Schweiz erfunden worden ist und mit all seinen charakteristischen Eigenschaften seit Beginn der vierziger Jahren auf dem freien Markte als

to be no easy matter, if we go by the opinion of John Delmonte, the well-known American glues and adhesives specialist, as expressed in his book "The Technology of Adhesives" [16], which was published as late as December 1946 – two years after the invention of ARALDITE on 26 November 1944. Discussing the American and British metal adhesives CYCLEWELD and REDUX, Delmonte states: "...Various others have been developed from time to time....It would seem that the goal for the ideal adhesive still lies ahead. Most of the strong metal-to-metal bonding types require...tooling costs of applying the correct heat and pressure for curing." Delmonte was still apparently unacquainted with CIBA Limited's ARALDITE "metal bonding types" that had already become available in the spring of 1946; and indeed these products did not involve any "tooling costs of applying the correct heat and pressure for curing". By his own standards, in fact, they already represented Delmonte's "goal for the ideal adhesive". There is even a certain irony in the fact that in the United States, where Delmonte was writing, the paint and varnish manufacturing firm of Devoe & Raynolds, then working in close collaboration with Shell Chemical Corporation, had been engaged since 1941 in the development of epoxy resins [17]. This work, however, was concentrated almost exclusively on the development of surface coating resins, and eventually resulted in the dominant position of these two companies in the field of protective surface coatings. No significant efforts were made at that time to develop the epoxies for other applications.

What is however astonishing, when seen in retrospect, is that the epoxies, known by their original name of ethoxyline resins, had already been invented in Switzerland in 1938 and had been freely available on the market since the early nineteen-forties in the form of a resin/curing agent mixture for the production of dentures. They were already being evaluated by synthetic resin specialists with the aim of exploring their potential applications, and their characteristic properties were there for everyone to see. Similarly the patents had been published already in 1940. Nevertheless their outstanding feature, their bonding function, which today is the most familiar characteristic of the resin and from the technological point of view the most important, i.e. the *combination* of properties already referred to, remained unrecognised from 1938 until 26 November 1944.

In the history of science and technology, as in history as a whole, there is no such thing as an identical repetition of events. Parallels abound, however. An obvious example is the substance known chemically as dichlorodiphenyltrichloroethane, which had been known for more than half a century before its specific function was realised and it was renamed D.D.T. All its properties – with one exception – had been discovered and were familiar to every chemist. This exception, its insecticidal action, is its most important property as regards both biological and commercial usefulness. It was this recognition of the special function of D.D.T. by Dr. Paul Müller of J. R. Geigy SA that represented the invention of the insecticide that is now a household name all over the world.

3. The origin of epoxy resins

The first definitely recorded identification and description of various phenolic, and in particular liquid, polyglycidyl ethers was given by Dr. Paul Schlack in 1934 in an I.G. Farben patent [18]. The actual subject of the patent was not, however, the series of compounds as such but their further utilisation for the production of polyamines [19]. A few years later the Swiss chemist Dr. Pierre Castan, working in the firm of Gebrüder de Trey AG in Zurich, manufacturers of dental products, followed up the work of Schlack; he reacted p-dihydroxydiphenyldimethylmethane (2,2-bis[4-hydroxyphenyl]propane), or bisphenol A, with epichlorohydrin [20]. The resulting fusible product, the ethoxyline resin, was then heated with phthalic anhydride, giving a hard, non-brittle substance of a light yellowish to brown colour that was described in the patent application as being

härtbare Harz/Härter-Mischung zur Herstellung zahnärztlicher Prothesen käuflich erhältlich war und in den Händen kompetenter Fachleute der Technik zwecks Prüfung von Anwendungsmöglichkeiten lag. Auch die Patente waren seit 1940 publiziert. Und dennoch blieb die neuartige, heute bekannteste und technisch interessanteste funktionelle Eigenschaft – die eingangs erwähnte Kombination – seit 1938 unerkannt. Die Erkenntnis erfolgte erst durch die ARALDIT-Funktionserfindung vom 26. 11. 44.

Wie in der allgemeinen Geschichte, so wiederholt sich auch in der Geschichte der Technik und Naturwissenschaften nichts in völlig identischer Weise. Es bilden sich aber immer wieder gewisse bemerkenswerte Parallelen. So liegt es nahe, an die DDT-Erfindung zu denken, bei welcher das chemische Produkt, das Dichlordiphenyltrichloräthan, seit mehr als einem halben Jahrhundert mit allen seinen Eigenschaften und Wirkungsmöglichkeiten vorlag. Nur eine Sache, eine allerdings für den technischen Fortschritt und die gewerbliche Brauchbarkeit sehr entscheidende, lag nicht vor: Die Erkenntnis der insektiziden Funktion. Diese Erkenntnis stellte damals den wesentlichen Inhalt der weltberühmten DDT-Erfindung des Geigy-Chemikers Dr. Paul Müller dar.

3. Die Entstehung der Äthoxylin-(Epoxy-)Harze

Die erste einwandfreie Darstellung und Identifizierung von verschiedenen phenolischen – besonders auch von flüssigen – Polyglycidyläthern beschrieb 1934 Dr. Paul Schlack in einem Patent von I.G.-Farben [18], dessen eigentlicher Gegenstand jedoch nicht die genannten Verbindungen, sondern deren Weiterverwendung zur Gewinnung von Polyaminen war [19]. Es war der Schweizer Chemiker Dr. Pierre Castan der Firma Gebrüder de Trey AG, einer Fabrik zur Herstellung zahnärztlicher Produkte, welcher während der 2. Hälfte der dreissiger Jahre diese Reaktion wieder aufgriff und Paradioxydiphenyldimethylmethan(4,4'-Dioxydiphenylpropan, auch kurz als Diomethan bezeichnet) mit Epichlorhydrin reagieren liess [20]. Das gewonnene schmelzbare Produkt, das Äthoxylin-(Epoxy-)Harz, erhitzte er weiter mit Phtalsäureanhydrid und gewann dadurch einen harten aber nicht spröden Stoff von leicht gelblicher bis brauner Farbe, den er in der Patentbeschreibung als unempfindlich gegen Wasser bis auf 80 °C und als temperaturbeständig bis auf 100 °C bezeichnete. Er beschrieb ihn als sehr gut bearbeitbar und erwähnte, dass das gehärtete Harz ausserordentlich gut an Glas, Porzellan und Metallen haftet, dass es ein guter Isolierstoff ist, dass man es in offenen Gefässen härten kann, ohne Gefahr zu laufen, poröse Stücke zu erhalten, und dass es sehr gut für Gießstücke verwendet werden kann. Er vermerkte die Verwendung als Presspulver, das Färben mit organischen und anorganischen Farbstoffen, wie auch das Versehen mit Füllstoffen und Plastifizierungsmitteln, wie Phtalsäureestern, Benzylbenzoat und Sipolin. Als Verwendung des gewonnenen Produktes wurde diejenige für gegossene und gepresste Gegenstände wie z. B. Galanteriewaren, elektrotechnische Artikel, Billardkugeln, Zahnersatzstücke angegeben. Es wurde auch auf die Verwendungsmöglichkeiten für Lackzwecke durch Einsatz des Harzes in gelöster Form hingewiesen.

Mit diesen Feststellungen war zum ersten Male ein gehärtetes Äthoxylin-(Epoxy-)Harz mit allen dieser Harzklasse eigenen und dieselbe später und bis heute kennzeichnenden Eigenschaften dargestellt und bekanntgemacht worden. Die diesbezügliche schweizerische Patentanmeldung erfolgte am 23. 8. 38 und wurde am 18. 11. 40 als Schweizer Patent Nr. 211116 druckschriftlich publiziert. Damit war der internationalen Technik und Chemie das Harz bekanntgemacht worden. Am 16. 6. 43 erfolgte noch eine weitere Anmeldung, welche die Härtung der Äthoxylin-(Epoxy-)Harze mit basischen Katalysatoren betraf (Schweizer Patent Nr. 236594).

Bei all diesem Geschehen war aber eines nicht erfolgt: Die Erkenntnis der eingangs erwähnten Kombination, d. h. der neuartigen höchstwertigen Bindefunktion. Dieses Erkennen musste noch einige Zeit auf sich warten lassen.

unaffected by water up to a temperature of 80 °C and resistant to temperatures of up to 100 °C. Castan described the new substance as being easily worked, and mentioned that the cured resin adhered outstandingly well to glass, porcelain, and metals ("ausserordentlich gut an Glas, Porzellan und Metallen haftet"), that it was a good insulator, that it could be cured in open vessels without risk of becoming porous, and that it was very suitable for producing shaped castings. He further pointed out that it could be used as a moulding powder, that it could be coloured with organic and inorganic colorants, and that it could also be used in combination with fillers and plasticisers such as phthalates, benzyl benzoate, and adipates (Sipolin). Applications suggested for the new resin included cast and moulded articles such as fancy goods, electrical equipment, billiard balls, and dentures. The patent also mentioned the possibility of using the resin in the form of a solution for surface coatings.

Dr. Castan's patent, in fact, was the first full and detailed description of a cured epoxy resin, complete with all the characteristic properties of this class. The patent application was filed on 23 August 1938 and the patent itself was published on 18 November 1940 as Swiss Patent No. 211,116. The development of the resin was in this manner announced to the world. A second patent application was filed on 16 June 1943, dealing with the curing of epoxy resins with basic catalysts, and was subsequently published as Swiss Patent No. 236,594. All this time, however, the special *combination* of properties of the resin went unrecognised, and the discovery of its unique bonding function still had to wait for some years.

4. Epoxy resins in prosthetic dentistry

Dr. Castan's primary concern was the use of his new resin for the production of dentures (cf. Swiss Patent No. 211,201, likewise applied for on 23 August 1938 by Gebrüder de Trey AG of Zurich); the resin was subsequently placed on the market in the early nineteen-forties as a casting resin under the trade name SELITROL for the use of dental practitioners. It was sold as a mixture ready for use and was formulated [21] and presented in such a manner as to compete with other polymers that had dominated the market for many years previously such as the methacrylates (Röhm and Haas, 1930; PALADON, Kultzer, 1936) [22], in that it could be utilised in accordance with the same standard techniques. Some years later it was withdrawn from the market, and its place was taken by other polymers that were simpler to use [23].

There is no doubt whatever that this SELITROL resin provided firm bonding of the teeth in the dentures. But its specific bonding function was still not recognised as such; indeed there was no need for a more powerful bond than could be provided by the competitive dental resins already in use [24]. Today the acrylics again dominate the market both in Europe and in the United States [25]; this is because their mechanical properties, including their bonding properties, are entirely adequate for the task that composites of this type are called upon to perform. The SELITROL resin had a tendency to stick to the moulds, like the acrylics: it stuck if anything even more, which was a disadvantage from the point of view of competing with the methacrylates. Indeed, its sticking tendency [26] necessitated a great deal of extra work, since the plaster mould had to be wrapped in tinfoil so as to cover any cracks or holes; otherwise the resin stuck to the plaster and white spots appeared on the casting [27]. It was thus a natural consequence that the unique property and essential function of the resin went unrecognised: only later, with the "invention of ARALDITE", was it fully realised that the epoxy resins were capable of performing a function of immeasurable importance for technology as a whole.

5. First attempts to investigate de Trey's SELITROL resin

A short time ago the electrical engineer Alfred Imhof, in an article entitled "On the borderline between electrical engineering and chemistry" ("An der Grenze zwischen Elektrotechnik

4. Die Äthoxylin-(Epoxy-)Harze als zahnärztliches Prothesenmaterial

Entsprechend der Aufgabe von Dr. Castan, Zahnersatzstücke wie Gaumenplatten, Gebisse usw., herzustellen (vgl. dazu das Schweizer Patent Nr. 211 201 der Firma Gebr. de Trey AG, Zürich, ebenfalls angemeldet am 23. 8. 38), wurde das Harz zu Beginn der vierziger Jahre als Giessharz unter der geschützten Wortmarke SELITROL zur Verfügung der Zahnärzte und Zahntechniker in den freien Handel gebracht. Es stellte eine gebrauchsfertige Mischung dar und war auf solche Art «formuliert» [21] und präsentiert, um es in Konkurrenz z. B. zu den seit Jahren vorhandenen und den Markt beherrschenden Polymerisaten aus Methacrylsäureestern (Röhm und Haas, 1930; Kultzers PALADON, 1936) [22] entsprechend den durch diese Harze u.a.m. eingebürgerten Verarbeitungsmethoden einsetzen zu können. Nach ein paar Jahren wurde es wegen der schwierigen Arbeitsweise durch einfacher handzuhabende Polymerisationsprodukte [23] ersetzt und vom Markte zurückgezogen.

Es besteht wohl kein Zweifel, dass dieses SELITROL-Harz an und für sich in den Gaumenplatten (Prothesen) die Zähne kraftschlüssig miteinander verbunden hat. Aber diese Funktion wurde weder erkannt noch in einem Masse benötigt [24], wie dieselbe nicht auch von einem andern Kunstharz hätte erfüllt werden können. Heute beherrschen in Europa wie in Nordamerika nach wie vor die Polymerisate aus Methacrylsäureestern das Feld [25]. Diese Harze tun dies, indem ihre Eigenschaften (auch Bindeeigenschaften) nach wie vor genügen, welche an die mechanischen Festigkeiten solcher Verbundkörper gestellt werden. Das SELITROL-Harz war ein «dichtendes» und «klebendes» Harz wie das Polymethacrylatharz. Es «klebte» vielleicht noch etwas mehr, was aber im Wettstreit mit den Polymethylmethacrylaten gar nicht gewünscht war. Das leidige Kleben [26] machte sogar notwendig, dass eine erhebliche Mehrarbeit bedingende Zinnfolie an die Gipsform «anrotiert» [27] werden musste, damit etwaige Risse und Löcher sorgfältig abgedeckt wurden, denn sonst «klebte» das Harz am Gips und es entstanden unerwünschte weisse Flecken am Werkstück. Und so ergab sich natürlicherweise, dass nicht erkannt wurde, was dann mittels der ARALDIT-Funktionserfindung zustande kam und für die Technik von grösster Bedeutung wurde.

5. Erste Überprüfungen von de Treys SELITROL-Harzen durch die Technik

Vor nicht langer Zeit berichtete Dipl.-Ing. Alfred Imhof [28] in einem Aufsatze «An der Grenze zwischen Elektrotechnik und Chemie» («Schweizerische Technische Zeitschrift», Sonderheft «Erfindungen und Erfinder», Nr. 52, Bern, 27. 12. 62), dass er in der 1. Hälfte der vierziger Jahre von seinem Zahnarzt kleinste Restbestände eines Kunstharzes erhalten habe, aus welchem er eine kleine Platte von etwa 4 × 6 × 0,4 cm goss – zu mehr reichte das Material nicht aus –, an der er in subtiler Weise tunlichst viele Qualitätsproben machte, deren Ergebnisse ihn ausserordentlich befriedigten und ihn sogleich nach den möglichen hochspannungstechnischen Anwendungsmöglichkeiten fahnden liessen. Es war deshalb nicht erstaunlich, dass er, nachdem auch ihn anlässlich der Schweizer Mustermesse im Mai 1946 am Stande der CIBA zum ersten Male die Kunde von dieser höchstwertigen Bindefunktion erreicht [29] und auch er von den «mannigfaltigen Metallklebstudien mit ARALDIT» – wie er schreibt – «gehört hatte», sich in der Folge von diesem Harz Muster kommen liess, um bei der Firma Moser-Glaser & Co. AG, Transformatorenfabrik, Muttenz bei Basel, «Giessversuche unter Einbettung von Spulen und Eisenkernen» vorzunehmen. Es sei festgehalten und gebührend vermerkt, dass damit die ersten ernsthaften Applikationen der Äthoxylin-(Epoxy-)Harze in der Hochspannungstechnik und eine weltweite Entwicklung auch auf diesem Gebiete ihren Anfang genommen hatten [30] (siehe Bild 2). Andere folgten nach (vgl. z. B. Gantenbein (sel.) und Koller der Maschinenfabrik Oerlikon (MFO)) und trugen ihrerseits [31] zur weiteren Entwicklung neue Erkenntnisse und Erfahrungen bei.

und Chemie", Schweizerische Technische Zeitschrift, special edition "Erfindungen und Erfinder", No. 52, Berne, 27 December 1962) [28], reported that in the early nineteen-forties he obtained from his dentist a minute quantity of a synthetic resin from which he made a small plate measuring about 4 by 6 by 0.4 cm – there was not enough material to make a larger one. By subtle methods he subjected this small plate to as many tests as he could; his results were highly satisfactory, and he decided to investigate the possible applications of the new material in high tension engineering. It is not surprising, therefore, that when he visited the CIBA stand at the Swiss Industries Fair in May 1946 and learnt about the remarkable bonding properties of the resin [29] and "the many metal bonding studies that had been carried out" – as he writes – he ordered samples with the intention of carrying out "casting trials for the potting of coils and iron cores" at his own works, the firm of Moser-Glaser & Co. AG, transformer manufacturers at Muttenz near Basle. These trials, it should be recorded, were the first serious attempts to apply epoxies to high tension electrical work, and marked the start of their subsequent successful application in this field all over the world [30] (Fig. 2). Other workers also carried out trials (e.g. Koller and the late A. Gantenbein of the Oerlikon Engineering Company [MFO]) and their results also contributed to further development [31].

In the meantime, namely since the end of November 1942, the Plastics Division of CIBA Limited in Basle had also been investigating the ethoxyline resins sold by Gebrüder de Trey AG: a team of workers were set the task not only of determining the physical (mechanical and electrical) properties of the resin but also of investigating its potential applications in the laboratory and on the shop floor. The resin was accordingly submitted to

Bild 2. Die erste fabrikatorische Verwirklichung unter Einsatz von Äthoxylin-(Epoxy-)Harzen auf dem Gebiete der Starkstromtechnik: 1 KV-Messwandler der Firma Moser-Glaser & Co. AG, Muttenz bei Basel.
(Aus: «Schweiz. Techn. Zeitschrift» 44/45, 6. November 1947. Vergleiche auch «Electrical Manufacturing», New York, Juli 1949, S. 78)

Fig. 2. The first product utilising epoxy resins in the field of power current engineering to be manufactured on an industrial scale: a one-kilovolt instrument transformer made by Moser-Glaser & Co. AG of Muttenz near Basle.
(From "Schweiz. Technische Zeitschrift", pp. 44–45, 6 November 1947; see also "Electrical Manufacturing", New York, July 1949, p. 78.)

Seit Ende November 1942 hatte aber auch die Kunststoffabteilung der CIBA in Basel die Prüfung dieser Äthoxylinharze der Firma de Trey an die Hand genommen und die zuständige fachmännische Instanz war von der Direktion beauftragt worden, nicht nur die physikalischen (mechanischen und elektrischen) Gütewerte festzustellen, sondern auch das Produkt in den Laboratorien und Werkstätten auf Applikationsmöglichkeiten zu untersuchen. In der Folge wurde das Harz auf seine mechanische Festigkeit (Messen der Standardwerte: Biegefestigkeit, Schlagbiegefestigkeit, Kerbschlagfestigkeit, Elastizitätsmodul), seine thermischen Eigenschaften (Messen des Standardwertes: Wärmefestigkeit nach Martens) und sein Verhalten gegenüber Wasser und mehr als 20 Chemikalien geprüft. CIBA liess auch beim Eidg. Amt für Mass und Gewicht in Bern und bei der AFIF in Zürich die elektrischen Eigenschaften (Standardwerte: Dielektrizitätskonstante und Verlustwinkel bei verschiedenen Temperaturen und Frequenzen, Durchgangs- und Oberflächenwiderstand) messen. Diese Messwerte bestätigten die bereits von de Trey offenbarte Eigenschaft eines guten Isolierstoffes, der in der Elektrotechnik z. B. auch in der Schwachstromindustrie Chancen haben könnte. Auch war von Anfang an erkannt gewesen, dass das giessbare Harz ohne Abspaltung flüchtiger Substanzen härtete und praktisch nichtschwindend war.

Niemand aber hatte die dem Harze innewohnende und dann am 26. 11. 44 erfinderisch festgestellte Bindefunktion erkannt.

6. Das ausserordentlich gute Haften an Glas, Porzellan und Metallen und das Nichterkennen der dann erfinderisch offenbarten Bindefunktion

Post festum – wie es bei Erfindungen etwa zu geschehen pflegt – mag mancher die an und für sich berechtigte Frage aufwerfen, warum denn damals in all den Jahren seit Schaffung dieser neuen Äthoxylin-(Epoxy-)Harzklasse im Jahre 1938 diese heute so berühmte Bindefunktion nicht eher erkannt worden war? Die Faktoren der hohen Adhäsion, der guten Kohäsion und des Nichtabspaltens von flüchtigen Substanzen während des Härtungsprozesses lagen von Anbeginn offen vor!

Vorerst darf darauf hingewiesen werden, dass auch in den USA das Äthoxylin-(Epoxy-)Harz seit der gleichen Zeitperiode bekannt gewesen war und dass dort bis zum Zeitpunkt des Eintreffens der Nachricht von der ARALDIT-Erfindung und des Erhaltens der ersten ARALDIT-Muster niemand die Erkenntnis dieser höchstwertigen Funktion eröffnet hatte. Viel wesentlicher ist aber doch wohl die Feststellung, dass in der Praxis der Begriff «ausserordentliches Haften» im Zusammenhange mit einer höchstwertigen Bindefunktion zwischen Metallen, Glas usw. technologisch noch nichts aussagte. Wie oft musste ein organisch-synthetisch forschender Chemiker harzartige Reaktionsprodukte, die bei einem meist ungewollten Reaktionsverlauf entstanden waren und die zu seinem Leidwesen «ausserordentlich fest» an der Glaswand seiner Gefässe hafteten, mit grosser Mühe wieder entfernen. Dieses ausserordentliche Haften war alles andere als erwünscht. Es stellte einen Nachteil dar und war in keiner Weise brauchbar [32]. Auch darf nicht übersehen werden, dass beim Gewinnen von Formkörpern in der Kunststofftechnik – sei es beim Pressen oder Giessen eines Gegenstandes – das Haften an der Metallform höchst unerwünscht ist. Unzählige Massnahmen sind versucht worden, um diesen Bindeeffekt zu verhindern. Auch bei den Äthoxylin-(Epoxy-)Harzen war von Anfang an das schwerwiegende Problem der Loslösung des Gießstückes aus der Form vorhanden [33] und so waren die Gedanken der Experimentatoren vorerst darauf gerichtet diesen Effekt zu verhindern, welcher später zu der entscheidenden Eigenschaft der neuen Harzklasse werden sollte. Der Laborant, der sich mit aller Kraft anstrengte sein gehärtetes Äthoxylinharzgussplättchen der Metallform zu entnehmen, wusste nicht, dass diese Formwände, zwischen welchen das Harz so leidig «klebte», mit einem «Kunstharzlot» höchster Güte und unübersehbarer Brauchbarkeit verbunden war.

searching tests to ascertain its mechanical properties (standard tests for flexural strength, impact strength, Izod notched impact strength, and modulus of elasticity), thermal properties (standard tests for Martens dimensional stability under heat), and its resistance to water and a score of different chemicals. CIBA also sent the resin to the Swiss Federal Bureau of Weights and Measures in Berne and to the A.F.I.F. in Zurich, where its electrical properties were measured (standard tests for dielectric constant and loss angle at various temperatures and frequencies, volume resistance and surface resistance). The findings confirmed Castan's original description of the resin as a good insulator that might prove useful in electrical engineering, for example in communications engineering. It had been recognised from the start that the coating resin cured without volatile substances being split off and that shrinkage was practically nil.

Still, however, nobody had recognised the specific "bonding function" of the resin, which was revealed only by the "invention of ARALDITE" on 26 November 1944.

6. Bonding function unrecognised despite outstanding adhesion to metals, ceramics and glass

Ex post facto, as is so often the case with discoveries and inventions, we may well wonder with some justification why the bonding function of the epoxies had not been discovered earlier, although these resins had been available and had been so closely studied ever since 1938: for all who had had to deal with this class of resins were fully aware of its "outstanding adhesion", its good cohesion, and the fact that no volatile substances were split off during the hardening process.

It has already been mentioned that Castan's ethoxyline resin had been equally well known in the United States for the same length of time; yet until the "invention of ARALDITE" was announced and the first samples arrived, the bonding function, by far the most significant feature of the resin, had remained unrecognised. Perhaps the factor most responsible was the fact that the phrase "outstanding adhesion to glass, ceramics, and metals" did not strike a responsive chord in the mind of the technologist. For it happened again and again that the experimental organic chemist obtained a resin-like substance, often unintentionally and as the product of a reaction carried out for some totally different purpose, which stuck stubbornly – with "outstanding adhesion" – to the sides of the test tube and could not be removed except with extreme difficulty. "Outstanding adhesion" under such circumstances was anything but desirable: it was a nuisance rather than an advantage, and was in no way useful [32]. Moreover, we should remember that when plastic parts are shaped, whether by casting or by moulding, adhesion to the metal mould is a feature that any plastics technologist regards as highly undesirable. Innumerable attempts have been made to prevent such sticking. This was of course also the case with the first ethoxyline resins: the problem of releasing the cast parts from the mould was a formidable one [33]. As a result, the efforts of resin chemists were initially concentrated on minimising this effect, which subsequently proved to be the most striking property of this new class of resins. The laboratory worker struggling to prise the cured ethoxyline casting from the metal mould did not stop to consider that the sides of the mould between which his resin was so firmly stuck were in fact bonded together by a "plastic solder" of superlative quality and tremendous practical utility.

Another case will illustrate the situation shortly before the invention of ARALDITE. In 1940 Professor Kurt Hans Meyer, in Vol. II of his "Die hochpolymeren Verbindungen" (one of the classical series of textbooks which, jointly with Professor H. Mark of New York, he edited under the general title of "Hochpolymere Chemie" [34]) stated on page 187 that the polyester resins (not, of course, epoxy resins) possessed the "striking property of adhering firmly to metals" ("ein auffallend starkes Adhäsionsvermögen auf Metallen"). Just as the "striking property of adhering firmly to metals" of the polyesters was not fol-

Schliesslich ist es auch notwendig, sich die folgende grundlegende technische Situation vor Augen zu halten: Prof. Dr. Kurt Hans Meyer hatte in seinem 1940 erschienenen Band II «Die Hochpolymeren Verbindungen» der zusammen mit Prof. Dr. H. Mark (New York) herausgegebenen klassischen Lehr- und Handbuchreihe «Hochpolymere Chemie» [34] auf Seite 187 festgestellt, dass Polyester (wohlverstanden nicht Äthoxylinharze!) ein «auffallend starkes Adhäsionsvermögen auf Metallen» besitzen. So wenig dieses «auffallend starke Adhäsionsvermögen auf Metallen» der Polyesterharze eine erwähnenswerte technische Folgerung in Richtung einer brauchbaren – den Äthoxylin-(Epoxy-)Harzen vergleichbaren – Metallbindefunktion verursachte, so wenig tat dies die Bekanntmachung des «ausserordentlich guten Haftens auf Glas, Porzellan und Metallen» der im de Trey-Patent beschriebenen Äthoxylin-(Epoxy-)Harze. Erst mit der ARALDIT-Funktionserfindung vom 26. 11. 44, welche die neue mit einem technischen Fortschritte verbundene Anwendung dieser seit Jahren bekannten Äthoxylin-(Epoxy-)Harze brachte, erfolgte jene Offenbarung, gemäss welcher das «ausserordentlich gute Haften» der de Trey-Harze nicht mehr in Parallele zu dem «auffallend starken Adhäsionsvermögen auf Metallen» der Polyester stand. Vom Zeitpunkt der bekanntgemachten Erfindung an wusste die Technik aller Applikationssparten, dass das «ausserordentlich gute Haften» im Falle der Äthoxylin-(Epoxy-)Harze eine höchstwertige, kraftübertragende Bindefunktion unter statischen wie dynamischen Gebrauchsbedingungen darstellte. Diese Funktion war in jenem Mai 1946 dem prüfenden Ingenieur an Hand überlappend ARALDIT-verklebter Blechstreifen mit Messwerten belegt worden. Von diesem Momente an war für die Allgemeinheit die Erkenntnis eröffnet, dass es von nun an möglich ist, nichtporöse Werkstoffe wie Metalle, Glas usw. – in allen ihren Formen [35] – als kraftübertragende Komponenten mittels eines Kunstharzes, ohne Benötigung eines spezifischen Pressdruckes, mit der durch die Funktionserfindung offenbarten Güte zu vereinen.

7. Das Entstehen der ARALDIT-Funktionserfindung

«Sonntag, den 26. November 1944 hatte Dr. Preiswerk, anlässlich eines privaten Besuches bei Dr. A. Gams (Direktor der CIBA und Leiter deren Kunststoffabteilung), mit diesem über das Problem der Metallverklebung diskutiert. Herr Dr. Gams hatte daraufhin vorgeschlagen, das de Trey-Harz zu versuchen.»

Mit diesen einfachen und kurzen Worten ist das entscheidende erfinderische Geschehen protokollarisch in der CIBA festgehalten worden. Wie war es zu diesem Ereignis gekommen?

Im Rahmen einer möglichst vielseitigen und vertieften Ausbildung als Chemiker war Dr. Preiswerk nach seinem Studienabschluss während der Jahre 1940 und 1941 Privatassistent bei dem bekannten Forscher auf dem Gebiete der Hochpolymeren Chemie, Prof. Kurt Hans Meyer [36] an der Universität Genf gewesen. Im Anschluss lernte er als Mitarbeiter des von Prof. Dr.-Ing. Alfred von Zeerleder geleiteten Forschungsinstitutes der Aluminium-Industrie-Aktiengesellschaft (AIAG) in Neuhausen am Rheinfall die Verwendung eines den Kunststoffen parallelen modernen Werkstoffes, des Leichtmetalles, in der Technik kennen. So war es naheliegend, dass nach einer solchen vorbereitenden [37] Schulung, die im übrigen selbständig und unabhängig konzipiert und durchgeführt worden war, sich bei ihm die Idee verdichtete, dass – trotz dem uralten und doch nie richtig erfüllten Wunsche der Technik nach einem organischen, kraftschlüssigen Bindemittel für nichtporöse Werkstoffe, wie Metalle, Glas usw. – die Prüfung dieser Frage beim heutigen Stande der hochpolymeren Chemie erneut an die Hand genommen werden sollte, im Sinne einer gegenseitigen Durchdringung von Kunststoff und Metall in der modernen Technik. Mit dieser anregenden Forderung von Dr. Preiswerk war die Initiative ergriffen und die Kommunikation geschaffen worden, zwischen einer dem technischen Fortschritte auf allen Gebieten der Technik im weitesten Ausmasse dienenden Aufgabe (dem Verbinden und Vereinigen der verschiedensten Werkstoffe, besonders solchen von nichtporöser Natur wie Metallen usw., mittels eines Kunstharzes in einem mechanisch und elektrisch lowed up with the aim of developing an adhesive usable for bonding metals, so the "outstandingly good adhesion to glass, porcelain, and metals" ("ausserordentlich guten Haftens auf...") of the ethoxyline resin described in the de Trey patent also failed to produce the reaction that, in retrospect, would have seemed so natural. It was only on 26 November 1944, the date of the "invention of ARALDITE", that the new field of application of the familiar de Trey resin was revealed and a whole new field of technological progress opened up: the "outstandingly good adhesion" of the epoxies had been rescued from the oblivion to which the "striking property of adhering firmly to metals" of the polyesters had been relegated. From the date of the announcement of the invention workers in all sectors of technology realised that the "outstandingly good adhesion" of the epoxy resins did in fact represent a new function of extreme importance and that these resins were capable of providing a bond with first-class mechanical properties under both static and dynamic conditions of loading. This unique "function" of the epoxy resins was demonstrated to engineers in May 1946 with the aid of data obtained by testing strips of sheet metal bonded together with ARALDITE. From this moment onwards the public was aware that it was now possible to bond together non-porous surfaces such as metal and glass, whatever their form [35], so as to produce high-quality composites capable of transmitting loads and that this could be done by using a synthetic resin that required no specific pressure for curing. The "function" of the epoxies had, in short, been discovered.

7. The "invention of ARALDITE" (the discovery of the bonding function)

"On Sunday 26 November 1944 Dr. Preiswerk, during the course of a private visit to Dr. A. Gams [the head of the CIBA Plastics Division], discussed the topic of bonding metals. Thereupon Dr. Gams suggested that the de Trey resin be investigated for this purpose."

This short note entered in the minutes of CIBA Limited in Basle records the "moment of truth" when ARALDITE was born. How did this come about?

After receiving a thorough training in the various branches of chemistry, Dr. Preiswerk, his studies completed, worked as assistant to Professor Kurt Hans Meyer, the well-known research worker in the field of macromolecular chemistry at the University of Geneva [36]. He afterwards worked under Professor Alfred von Zeerleder at the Research Institute of Swiss Aluminium Industry Limited (AIAG) at Neuhausen, where he became acquainted with the technology of light alloys – another modern material of which the development has in many respects been parallel to that of plastics. Thus with a solid grounding in these two separate fields of knowledge [37] (Preiswerk had in fact independently planned and carried out his own educational programme), it was natural that his thoughts should turn once again to the old but as yet never satisfactorily solved problem of finding an organic, powerful bonding agent for non-porous materials such as metal and glass. The chemistry of high polymers had in recent years made great strides, and it was obvious that technology as a whole would gain enormous benefit from a marriage of plastics and metals. He took up the idea in earnest, and the necessary link was thus provided between the task to be performed on the one hand and the available tool on the other: the task, of immeasurable importance to all sectors of technology, being that of bonding and joining all kinds of materials, especially non-porous ones such as metals, so as to provide a composite material of optimum mechanical and electrical properties; and the tool being the ethoxyline resin whose synthesis had been announced four years previously and which was readily obtainable on the market. The new bonding function had been revealed and the "invention of ARALDITE" had become reality. Subsequent developments were to occupy the attention of technologists throughout the world, and were unforeseeable in their magnitude and scope.

hochwertigen Verband) und dem seit 4 Jahren öffentlich bekanntgemachten und allgemein käuflich erhältlichen Mittel, welches diese Aufgabe erfüllen konnte. Die ARALDIT-Funktionserfindung war zustande gekommen und das Tor zu einer unübersehbaren und weltweiten Entwicklung geöffnet.

Überprüfende Versuche mit Mustern dieses de Trey-Harzes auf dem Gebiete der Leichtmetalle, welche dann gleich von Dr. Preiswerk bei Prof. von Zeerleder beantragt worden waren und von ihm dann durchgeführt werden durften [38], bestätigten auf eindrückliche und vielseitige Weise (Festigkeitsmessungen bei statischer und dynamischer [39] Beanspruchung, Verbinden verschiedenster Werkstoffe, Herstellung von Laminaten usw.) die neue überragende Funktion. Ein in der Folge sehr direkter und vertrauensvoller Kontakt zwischen CIBA und AIAG bewirkte, dass dieses neue Hilfsmittel der Technik, dessen neue sensationelle Eigenschaften nur allzu leicht Anlass zu Fehlentwicklungen gegeben hätten, von den Verbrauchern der Leichtmetalle gleich von Anfang an verantwortungsbewusst und mit den notwendigen fachmännischen Überlegungen eingesetzt wurden. Auf diese Art wurde verhindert, dass Fehlanwendungen, welche der Sache der Kunstharze wie auch der Leichtmetalle geschadet hätten, nicht eintreten konnten und in der Folge praktisch auch nicht eintraten. Die Entwicklung wurde so gleich von Anfang an in die richtige auf einen langfristigen Erfolg ausgerichtete Bahn geleitet und auf ihr beibehalten [40]. Im Herbst 1945 nahm Dr. Preiswerk seine Arbeit in der CIBA auf. Sie dauerte bis zum Jahre 1957.

Am 13. Juli 1945 meldete CIBA in der Schweiz für diese Funktionserfindung [41] das grundlegende Patent an. Während des Winters 1949/50 wurde vom Beschwerdesenat des Niederländischen Patentamtes [42] und im Frühling 1956 vom Beschwerdesenat des Deutschen Patentamtes das Patent auf die gegenüber der Prioritätsanmeldung praktisch unveränderten Erfindungsansprüche erteilt. Die Erfindung war übrigens in über 50 Patentländern angemeldet worden.

Im April 1946 erfolgte anlässlich der schon eingangs erwähnten Schweizer Mustermesse die grundlegende Bekanntmachung Preiswerk und von Zeerleder [5] an die Technik über Applikations- und Prüfresultate. Eine offizielle, an der Eidg. Materialprüfungs- und Versuchsanstalt Zürich durchgeführte eingehende Prüfung konnte dann in schönster Weise diese Ergebnisse bestätigen [43]. Sie standen auch in Übereinstimmung mit den Erfahrungen, über welche die Technik aus allen Ländern [44] und Anwendungsgebieten in der Folge in immer steigendem Masse berichten konnte [45].

Das erste wohlgeprüfte Verkaufsprodukt, für welches die geschützte Wortmarke ARALDIT ausgewählt worden war und welches als sog. «ARALDIT-Bindemittel Typ I» in Form von «Pulver» und «Stangen» (in den Farben «nature» und «silber») im Mai 1946 den Mustermessebesuchern angeboten wurde [46], stellte eine einfache, leicht handhabbare und lagerfeste Präsentation dar. Sie erleichterte der Technik, die offenbarte, höchstwertige Bindefunktion mittels selbsthergestellter Prüfmuster zu messen und Applikationsversuche an die Hand zu nehmen. Dieses so «formulierte» Produkt – nun bald 20 Jahre alt und bis heute praktisch unverändert – war der erste Repräsentant eines technischen Verkaufsproduktes dieser Äthoxylinharzklasse, deren erfinderisch erkannte, neuartige und höchstwertige Bindefunktion – jene «Kombination» der seit 1938 bekannten Eigenschaften (Adhäsion, Kohäsion, kein Abspalten flüchtiger Substanzen) – sie auf raschestem Wege in alle Gebiete der Technik eintreten liess und damit ein neues Kapitel in der Geschichte der modernen Werkstofftechnik eröffnete.

8. Die weitere Bewährung der Äthoxylin-(Epoxy-)Harze

Von Anfang an waren durch CIBA neue Harze mit verschiedenen Polyphenolen geprüft worden, die aber zeigten, dass die ursprünglichen Äthoxylin-(Epoxy-)Harze (Äthylenoxydderivate eines Phenoles, insbesondere des 4,4'-Dioxy-diphenyl-2,2'-propans) die brauchbare Grundlage auch für diese neue Applikation

Dr. Preiswerk immediately proposed to Professor von Zeerleder that he carry out tests with samples of the de Trey resin to investigate its applicability in the field of light alloys. Tests were sanctioned [38], and the newly realised "function" of the epoxies was amply and impressively confirmed (mechanical bonding strength under static and dynamic [39] loading, bonding of a wide range of materials, preparation of laminates, etc.). Afterwards close contact between AIAG and CIBA was successfully established and prevented the mistakes that might have resulted from rash and possibly incorrect application of the new and sensational knowledge that had been acquired. The new servant of modern technology was thus from the very outset put to use by light alloy users who were fully aware of its true capabilities, and failure, which would have been detrimental to plastics and to the aluminium industry alike, was successfully avoided. In this manner development proceeded along the right lines from the very beginning, and ultimate success was assured [40]. In 1945 Dr. Preiswerk continued his work in CIBA Limited, where he worked until 1957.

On 13 July 1945 CIBA filed the basic patent application in respect of the invention of the function [1] [41]. During the winter of 1949/1950 the Appeals Tribunal of the Netherlands Patent Office [42], and in the spring of 1956 the Appeals Tribunal of the German Patent Office granted the patent in terms virtually identical with those of the original priority application. Patents were in addition applied for in more than fifty countries.

In April 1946, on the occasion of the Swiss Industries Fair already mentioned, the original paper by Preiswerk and von Zeerleder announcing the results of the application trials and tests was published [5]. Detailed testing carried out at the official level by the Swiss Federal Laboratory for Testing Materials and Research in Zurich provided confirmation of the results published [43]. They were further confirmed by an increasing number of reports covering numerous sectors of technology that soon began to arrive from many countries [44, 45].

The first CIBA epoxy to be thoroughly tested and prepared for commercial exploitation was assigned the trade name "ARALDITE" and, under the designation "ARALDITE Bonding Resin Type 1", was presented in the form of powder and sticks (in the colours "natural" and "silver") at the May 1946 Swiss Industries Fair [46]; it was a simple product, easy to manipulate, and with a good storage life. It was particularly convenient for producing test specimens so that the newly revealed bonding function could be tested and application trials initiated. This formulation – now almost twenty years old and still virtually unchanged today – was the first commercial representative of the class of epoxy or ethoxyline resins, and its novel, newly invented bonding function, the *combination* of properties (adhesion, cohesion, no splitting off of volatile substances on curing) already known since 1938, ensured its rapid penetration into all sectors of industry and thus opened up a new chapter in the history of modern materials technology.

8. Further experience with epoxy resins

From the very outset CIBA Limited tested numerous other new epoxy resins made with various polyphenols, but it became apparent that the original ethoxyline or epoxy resins – ethylene oxide derivatives of a diphenol, and particularly of 2,2-bis(4-hydroxyphenyl)propane, or bisphenol A – were the practical basis of the new function. This is still true today. The most suitable hardeners were identified without much difficulty by subjecting various chemical substances to simple tests. It proved a relatively simple matter to find the appropriate hardening agent for any given application and to establish the ideal operating conditions – quantity, curing temperature, curing time, etc. Similarly it was not difficult to select appropriate additives such as fillers, colorants, resins, and plasticisers from the range already available for the modification of other plastics. In this manner the new material, the epoxy resins with their newly recognised

---
[1] See footnote 4 on page 4.

blieben und es auch heute noch sind. Bei den Härtungsmitteln liess sich die Brauchbarkeit der verschiedenen chemischen Stoffklassen durch einen einfachen Versuch feststellen. Dem Fachmann gelang es ohne grosse Schwierigkeiten, das für den jeweiligen Zweck geeignete Härtungmittel auszuwählen und die besten Anwendungsbedingungen, wie Menge, Härtungstemperatur und Härtungsdauer festzulegen. In vergleichbarer Weise liessen sich Zuschlagsstoffe wie Füllmittel, farbgebundene Stoffe, Harze, Plastifizierungsmittel u. dgl. aus dem üblicherweise für die Modifikation von Kunststoffen bekannten Sortiment auswählen. Es waren dies jene Massnahmen, wie sie unter dem Begriffe «Formulieren» [47] bekannt sind und welche gestatten, das neuartige Arbeitsmittel, wie es mit seiner Funktion als kraftschlüssiger Metallbinder in den Äthoxylin-(Epoxy-)Harzen erkannt und offenbart worden war, den allerverschiedensten speziellen Anforderungen anzupassen. Auch waren von Anfang an Applikationen entwickelt und bekanntgemacht worden (siehe Bild 3), welche aufzeigen sollten, dass es nicht nur möglich ist, das Harz für die verschiedensten Methoden des «Anortbringens» anzupassen (Harz in Form von Tabletten, Verarbeitung mit der Spritzpistole usw.), sondern dass auch die maschinelle Serienverarbeitung ein anzustrebendes und mögliches Fertigungsverfahren für die verschiedensten Anwendungsfälle darstellt bei welchen Äthoxylin-(Epoxy-)Harze zum Einsatz kommen.

Ende der vierziger Jahre sich anschliessende Untersuchungen (Dr. W. Fisch/CIBA) über den Chemismus bei der Herstellung der Äthoxylinharze (Umsetzung von Bisphenol und Epichlorhydrin) und über die Härtung der Äthoxylinharze durch Polycarbonsäureanhydride schufen sehr beachtete und wertvolle Einblicke in den Reaktionsverlauf [48].

In jüngster Zeit wurde auch die Epoxydierung ungesättigter Verbindungen mit Kohlenstoff-Doppelbindungen in Angriff genommen [49], welche speziell erlaubte, in Richtung der Warmfestigkeit dieser die Eigenschaft einer höchstwertigen Bindefunktion beinhaltenden Äthoxylin-(Epoxy-)Harze [50] Fortschritte zu erzielen.

9. Namen, Bezeichnungen und Begriffe im Zeichen des technischen Fortschrittes

Mit dem Fortschreiten und Entwickeln der Technik ist es unvermeidlich, dass die in ihrem Bereiche verwendeten Namen, Bezeichnungen und Begriffe einem Wandel unterliegen können [51].

Die Äthoxylin-(Epoxy-)Harze (als «Rohharze» oder als «formulierte Produkte») waren vom Zeitpunkte der offenbarten Funktionserfindung an das neuartige, frei fliessende [52] «Kunstharzlot». Ihm wohnte [53] die wesentliche Eigenschaft inne, dass, sobald es zwischen den zu verbindenden Flächen vornehmlich nichtporöser Werkstoffe wie Metalle, Glas usw. zur Härtung gebracht wurde, die Werkstoffkomponenten – gleich welcher Gestaltung – in einem höchstwertigen Verbande vereinigt wurden. Der Fugendicke waren dabei grundsätzlich keine Grenzen mehr gesetzt, wie dies bei den bisherigen unter der Notwendigkeit einer spezifischen Druckanwendung stehenden «konventionellen» [54] Klebstoffen der Fall gewesen war [55]. Der Begriff «Klebstoffe» («adhesives») musste offensichtlich eine Weiterung erfahren, und so kam es nicht von ungefähr, dass im Jahre 1954 auch in den Deutschen Normen 16920/16921 (Juni/Juli 1954), «Klebstoffe, Richtlinien für die Einteilung», «Klebstoffe, Klebstoff-Verarbeitung, Begriffe» [56], die Begriffe [57] der «wirklichen Haftfestigkeitsforderung»/«der besonderen Anforderung an die Haftfestigkeit» und der «plastisch formbaren Klebstoffe, ... die auch zum Füllen dickerer [58] Klebefugen dienen können», Erwähnung gefunden haben. Ob nun in der Folge die Bezeichnungen dieses «Kunstharzlotes» auf Äthoxylin-(Epoxy-)Harzbasis das eine Mal mehr den Verarbeitungscharakter («Giessharz» = flüssiges, giessbares Harz [59]; «Imprägnierharz» = Harz zum Imprägnieren usw.) beinhaltete oder das andere Mal mehr auf das Anwendungsgebiet («Laminierharz» = Herstellung von Laminaten; «Werkzeugharz» = Herstellung von Werkzeugen usw.) hinwies (wobei auch solche Bezeichnungen sich überschneiden und ändern können),

Bild 3. Das «Anortbringen» des Äthoxylin-(Epoxy-)Harzes – zum Beispiel in Tablettenform zum «Verlöten» von Keramik und Messing – war schon frühzeitig ein Fingerzeig in Richtung der automatisierten Verarbeitung. Nicht nur galt es das Harzprodukt den Fertigungsverfahren anzupassen, sondern sogar eigene maschinelle Verarbeitungseinrichtungen zu entwickeln.
(Aus: «Electrical Manufacturing», New York, Juli 1949, S. 80, Fig. c)

Fig. 3. The various methods used for putting the epoxy adhesive "in place" – e.g. in the form of pellets for "soldering" ceramic and brass parts – was an early indication of automatic techniques that were subsequently to be developed. Not only was the resin adapted for existing production methods, but new methods were also evolved to suit the special properties of the resin.
(From "Electrical Manufacturing", New York, July 1949, p. 80, Fig. c.)

ability to function as high-strength metal bonding agents, could be "formulated" [47] so as to answer a wide range of specialised requirements. Applications were also developed and announced from the very beginning (see Fig. 3), whereby it was demonstrated not only that the resin could be adapted to the various existing methods of applying the bonding agent (in the form of pellets, with a spray gun, etc.) but also that continuous operation with automatic machines was a practical and desirable possibility in many of the fields of application where epoxy resins were used. Towards the end of the nineteen-forties Dr. W. Fisch of CIBA carried out a series of investigations into the chemical processes involved in the preparation of ethoxyline resins (reaction between bisphenol A and epichlorohydrin), and also did valuable work on the mechanism by which epoxies are cured with polycarboxylic acid anhydrides [48].

More recently research has also been done on the epoxidation of unsaturated compounds with carbon-to-carbon double bonds [49]; the result has been an improvement in the heat distortion temperature [50].

9. Terminology

As science and technology progress and develop it is inevitable that the nomenclature and terminology used undergo modification [51].

Whether in the form of "pure resin" or "formulated product", the epoxy resins possessed from the outset, that is to say from the date that the invention of their bonding function was first disclosed, the character of a free-flowing [52] plastic material reminiscent of metal solder in application. Their special property consisted in the fact that as soon as they were cured between the surfaces of two components – notably non-porous ones such as metals, glass, etc. – the two components, whatever their form, were united together in a composite of outstandingly high quality [53]. Whereas the more conventional [54] adhesives required the application of a specific bonding pressure [55], the thickness of the gap between epoxy-bonded components is largely immaterial, being limited only by practical considerations. The term "adhesives" obviously required clarification. Thus in 1954 the German Industrial Standards (DIN Standards 16920/16921 dated June/July 1954: "Adhesives, classification of" and "Adhesives and their application, definition of terms") made specific reference to "the real concept of bond strength" and "the special concept of bond strength" ("wirkliche Haftfestigkeitsforderung" and "besondere Anforderung an die Haftfestigkeit") and to "plastic adhesives that may also be used for the filling of larger gaps" ("plastisch formbare Klebstoffe, ... die auch zum Füllen dickerer Klebefugen dienen können") [56, 57, 58].

Bild 4. SNCASO's (heute Sud-Aviation) «Vautour»! Die kraftübertragende Flügelbeplankung des «Vautour» besteht aus überlangen, starkwandigen Blechen (von nach aussen abnehmendem Querschnitt) und dicken angeklebten Längsprofilen. Erstmals wurde hierbei das Metallkleben an lebenswichtigen Teilen erprobt. Das verwendete Bindemittel war ein Äthoxylin-(Epoxy-)Harz (ARALDIT Typ 1 der CIBA).
(Aus: «French Use New Bonding Method on Thriple Threat Vautour», W. P. Moser, «Aviation Age», New York, Dez. 1954, S. 38ff.)

Fig. 4. The "Vautour" tactical bomber made by S.N.C.A.S.O. (now Sud-Aviation). The stressed-skin wing of the "Vautour" consists of long heavy-gauge metal sheets (of diminishing thickness towards the wing tips) with thick longitudinal stiffeners. This was the first use of epoxy resins for bonding vital aircraft parts. The metal-bonding agent used was ARALDITE Type I, made by CIBA Limited.
(From "French Use New Bonding Method on Triple Threat Vautour" by W. P. Moser, "Aviation Age", New York, December 1964, pp. 38 ff.)

Bild 5. Lord Ismay besichtigt in seiner Eigenschaft als Generalsekretär der NATO das französische Flugzeugwerk SNCASO, wo ihm die Produktion des französischen Kampfflugzeuges «Vautour» vorgeführt wird. Mit seiner linken Hand betastet er die mittels Äthoxylin-(Epoxy-)Harz (ARALDIT Typ 1) verbundenen Flügelpartien.
(Aus: «Revue-SNCASO», Nouvelle Série, N. 3, Paris, Oktober 1953)

Fig. 5. Lord Ismay, then Secretary General of N.A.T.O., visiting the works of S.N.C.A.S.O., the French aircraft manufacturers, where he saw the "Vautour" tactical bomber in production. He is seen examining the wing components bonded with epoxy resin (ARALDITE Type I).
(From "Revue-SNCASO", Nouvelle Série, No. 3, Paris, October 1953.)

so wusste der Verbraucher, dass er sich bei dessen Einsatz stets mit grösstem Vorteil der im Vordergrunde stehenden erfinderisch offenbarten Bindefunktion zur Schaffung eines wertvollen Gebrauchswertes seiner Fabrikate bedienen konnte. Unter diesem Aspekt handelte tatsächlich auch die Technik (vgl. Abschnitt 10, nachstehend) [4, 84].

In diesem Zusammenhange überrascht auch nicht, dass der Begriff «Giessharz», der vor ungefähr 20 Jahren von der massgebenden Fachwelt vornehmlich noch mit den Phenolharzen («Edelkunstharzen») zur Herstellung von Formstücken (oder doch wenigstens Formrohlingen, die durch Drechseln oder andere Verfahren weiter verarbeitet werden können) identifiziert wurde [60], heute richtigerweise vorerst einmal allein mit der

In the case of the new epoxy resin-based "plastic solder" the product could be classified in such a manner as to reflect either its mode of application ("casting resins", i.e. liquid, pourable resins, "impregnating resins", etc. [59]) or its field of application ("laminating resin", "tooling resin", etc.). Although most of these terms overlap to some extent and are constantly subject to modification, the user was nevertheless aware that when applying the new resins he was in a position to utilise to the full their novel bonding function in such a manner as to produce a composite of outstanding performance. And these were in fact the general principles which guided industrial technologists in pursuing their individual and specific lines of development (see section 10 below) [4, 84].

Bild 6. CONVAIR's B58 «Hustler» wird auch als «the flying glue line» bezeichnet. Von der Bindekraft der Äthoxylin-(Epoxy-)Harze – ca. ½ Tonne pro Apparat – wird in der Konstruktion grösster Nutzen gezogen, um Leichtmetallblechstücke, Glasfasern und «Honigwaben»-Material zu kombinieren.

Fig. 6. The Convair B58 "Hustler" has been described as "the flying glue line". The design of the bomber makes full use of the bonding strength of epoxy resins – approximately half a ton per aircraft – in combinations of light alloy sheet, glass fibre, and honeycomb material.
(From an advertisement of Shell Chemical Corp. in "Aeronautical Engineering Review", April 1958, p. 75.)

Eigenschaft eines flüssigen oder schmelzbaren und dann härtbaren Harzes [61] gekennzeichnet wird und auch unter dem Titel «Imprägnierharz» [62] oder unter andern Bezeichnungen Verwendung findet.

Dass die Eigenschaften des Nichtabspaltens von flüchtigen Substanzen und des geringen Schwundes beim Härten, welche sich neben anderem als bestimmend günstige Voraussetzung für die Funktion des neuartigen «Kunstharzlotes» erwiesen, auch etwa für die Herstellung grosser, blasenfreier und nur wenig rissanfälliger Formstücke (Werkstücke, Formstoffe) von Bedeutung [63] waren, sei nebenbei vermerkt.

Die Möglichkeit aber, im gleichen Zuge sowohl eine höchstwertige Verbindung zwischen den Flächen zweier Werkstoffkomponenten (Metallen, Glas usw.) wie auch gleichzeitig ein hochwertiges Formstück – wenn nötig erheblichen Ausmasses – zu realisieren, wurde für die Äthoxylin-(Epoxy-)Harze typisch.

10. Die Bedeutung der Bindefunktion der Äthoxylin-(Epoxy-)Harze in der Praxis der Technik

Im Rahmen der Zielsetzung, bei welcher die technische Bewährung, auch unter anspruchvollsten Bedingungen, als alleiniger Ausdruck eines echten und endgültigen Erfolges gewertet wird, waren es Meilensteine in der Entwicklung als die ersten Vertreter der Flugkörperindustrie von der Qualität der nun bei industriell

Bild 7. Die Äthoxylin-(Epoxy-)Harze dienen nicht nur zum Dichthalten der Flügelkasten von CONVAIR's «Coronado», sondern sie bringen mit einer gewonnenen Scherfestigkeit von 2,8 kg/mm² eine zusätzliche Gesamtfestigkeit des Flügels.
(Aus: Publ. «Die Strahlflugzeuge der SWISSAIR», Dep. Technik, Ing. Abt. der SWISSAIR, Jan. 60, Abschnitt CV 600)

Fig. 7. In the Convair "Coronado" epoxy resins are not only used for bonding the wing box spars but, by providing a shear strength of 2.8 kg/mm², they add to the total strength of the wing.
(From "Die Strahlflugzeuge der Swissair", published by the Technical Department of the Engineering Division of Swissair, January 1960, section on "CV 600".)

Bild 8. Ungefähr 25 % des Gewichtes der hochbeanspruchten Konstruktionsteile wie Rumpf, Leitwerk und Steuerorgane dieser Fliegerabwehrrakete OERLIKON bestehen aus Äthoxylin-(Epoxy-)Harz-Bindemitteln (ARALDIT).
(Aus: «Aviation Magazine», Paris, 1. Juni 1958, S. 35)

Fig. 8. Approximately 25 per cent of the weight of the heavy-duty structural components of the Oerlikon anti-aircraft missile, including the casing, tail assembly, and control elements, is accounted for by epoxy resin-based bonding agents (ARALDITE).
(From "Aviation Magazine", Paris, 1 June 1958, p. 35.)

Since terminology is constantly subject to revision, it is not surprising that the term "casting resin" ("Edelkunstharz"), which some twenty years ago was regarded by most plastics specialists as being for practical purposes identical with the term "phenolic resin" (for the production of shaped castings or at any rate of rough castings that could be further worked by turning or other methods [60]), it is nowadays correctly used only with reference to a liquid or liquefiable, curable resin [61] that may also be classed as an "impregnating resin" [62] or even under some different designation.

In this context it may also be pointed out that the epoxy resins do not release volatile substances on curing and that their curing shrinkage is minimal, which in combination with their other properties represents the essential requirement for their function as a novel "plastic solder" and which is also of significant value to the producer of larger shaped castings (mouldings, cast resin) that have to be free from entrapped bubbles and exhibit only a minimum tendency to crack [63].

However, the possibility of producing a bond of high strength between the adjacent surfaces of two components (metal, glass, etc.) and at one and the same time of producing a high-grade casting, if necessary of considerable dimensions, became recognised as typical of the epoxies.

10. The importance of the bonding function in technological practice

Any new technological development must pass the strictest of tests before it can be accepted as a true advance: it must satisfy all the essential requirements, even the most stringent, that are posed in actual practice. For this reason one of the most important milestones in the history of the epoxy resins was when, for the first time, representatives of the aircraft industry were able to report on the quality of the aircraft and missile components (see Figs. 4, 5, 6, and 7 and Figs. 8 and 9 respectively) [64, 65] that had been produced on an industrial scale and in which the bonding properties of ARALDITE were systematically exploited. It was inevitable that space vehicle constructors also would eventually turn to the new bonding agents (see Fig. 10), and further development took undreamt-of turns the outcome of which still cannot be foreseen [66, 67, 68]. Scarcely less important was the influence of the epoxies on virtually all sectors of technology,

gefertigten Flugzeug- [64] (siehe Bild 4, 5, 6 und 7) und Raketenteilen [65] (siehe Bild 8 und 9) in grossem Ausmasse planmässig eingesetzten Bindefunktion der Äthoxylin-(Epoxy-)Harze berichten konnten. Die Folge war eine Weiterentwicklung erstaunlichen [66] und noch unübersehbaren Ausmasses [67], von welcher nun auch zwangsläufig der Raumkörperbau [68] profitierte (siehe Bild 10). Nicht minder war auch die Ausstrahlung auf praktisch alle weiteren Gebiete der Technik, insbesondere auch der Elektrotechnik. Es ist deshalb begreiflich, wenn z. B. anlässlich des europäischen Kongresses eines führenden internat. Elektrokonzerns ein Mitglied aus den eigenen Reihen den Teilnehmern die Zuverlässigkeit («reliability») der Bindefunktion,

and particularly on electrical engineering. Thus at the European congress of a leading international electrical engineering concern one of the delegates reported on the reliability of the adhesive bond given by ARALDITE, of fundamental importance in all the numerous composites made with epoxy resins that are used in the production of electrical equipment; the speaker illustrated his comments with the aid of a diagram (see Fig. 11) showing the shear strength values determined on an industrially produced aircraft wing [69]. Wherever in heavy (see Figs. 12 and 13) or light current engineering and electronics (see Fig. 14) the bonding properties of the new "plastic solder of high insulating power" [70] were utilised for the joining of different materials,

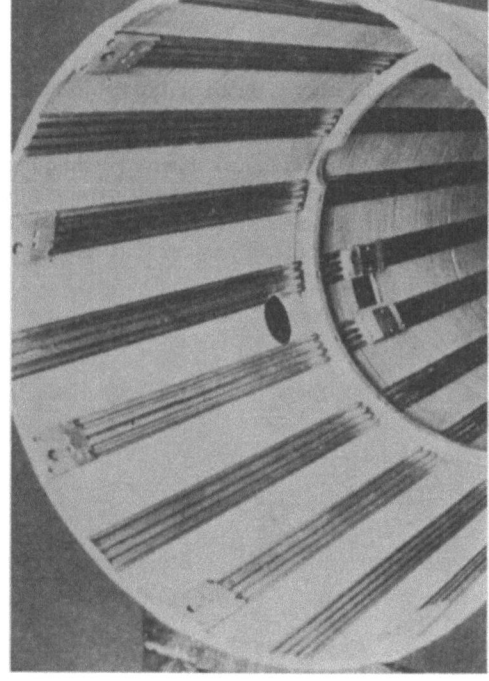

Bild 9. Die Rumpfschale der OERLIKON-Fliegerabwehrrakete wurde mittels des heute zu grösster Bedeutung gewordenen «Wickelprozesses» («filament winding») aus einem Leichtmetall-Wickelrohr gefertigt. Die einzelnen Wickellagen und die Rumpf-Längsprofile wurden mittels Äthoxylin-(Epoxy-)Harz (ARALDIT Typ 1) verbunden. «Es sind Metallverbindungen möglich, die eher unter den Begriff «Vergiessen» als unter dem herkömmlichen Begriff von «Verkleben» einzuordnen sind», schrieb Ing. Stencel in seinem Bericht. («Aluminium Suisse», Zürich, Sept. 1956, S. 149ff. «Neue Leichtbauweisen für die Zelle der leitstrahlgesteuerten OERLIKON-Fliegerabwehrrakete»)

Der Wickelprozess («filament winding») erlaubt unter Verwendung von Bändern oder Fasern aus Metall, Glas oder anderen Materialien und Äthoxylin-(Epoxy-)Harzen die Herstellung von Raketenmotorgehäusen («metal strip wound rocket motor cases»), Gewehr- und Geschützläufen («gun barrels»), Unterwasser-Apparaten («submarine components»), elektrischen Löschkammern («circuit breakers»), Hochspannungsrohren («high voltage tubes»), Federstücken («springs») und Druckgefässen («pressure vessels») usw. J. J. Sewell (NARMCO Mat. Div. Telecomputing Corp.) definierte kürzlich: «Filament-wound structures are classified as «composite structures» in that different materials are formed into an integrated mass for load-transmitting purposes.» («Mod. Plast.», New York, Febr. 62, S. 77)

Fig. 9. The main casing of the Oerlikon anti-aircraft missile was produced by the technique of filament winding, a process that is today of fundamental importance in the aircraft industry. A tubular profile of light alloy was used, and the individual windings were bonded to the longitudinal members with an epoxy resin (ARALDITE Type I). "Metals can be bonded by processes that are better described by the term 'casting' than by 'bonding' in the more conventional sense of the word", is the verdict of F. B. Stencel in his report "New lightweight construction techniques for beam-riding Oerlikon anti-aircraft missile" in "Aluminium Suisse", Zurich, September 1956, pp. 149 ff.

The filament winding process utilises epoxy resins with strips or fibres of metal, glass, or other materials, and may be employed for the construction of metal strip-wound rocket motor cases, gun barrels, submarine components, circuit breakers, high voltage tubes, springs, pressure vessels, etc. J. J. Sewell (NARMCO Mat. Div. Telecomputing Corp.) recently summarised: "Filament-wound structures are classified as 'composite structures' in that different materials are formed into an integrated mass for load-transmitting purposes". ("Modern Plastics", New York, February 1962, p. 77.)

Bild 10. Der Courier 1-B-Satellit ist zusammengesetzt aus zwei Hälften von glasfaserverstärktem Äthoxylin-(Epoxy-)Harz, mit welchem gleichzeitig 20000 Silizium-Sonnenzellen verbunden sind. Die Bindekraft des Harzes ermöglichte eine vollständig homogene Verbindung zwischen dem Metall, dem Glasgewebe und dem «Bienenwaben»-Material an kritisch beanspruchten Partien des Satelliten. Das ausgewuchtete Modell wird soeben auf einem Vibrationstisch geprüft, wo die Schwingungen simuliert werden, die während der Antriebsphase der Trägerrakete auftreten.
(Aus: «Flugkörper», 1960, Heft 12, S. 380)

Fig. 10. The "Courier 1-B" satellite is composed of two hemispheres of glass fibre-reinforced epoxy resin to which 20,000 silicon solar cells are also bonded. The high adhesive strength of the resin made possible a completely homogeneous bond between the metal, glass cloth, and honeycomb material in parts of the satellite subjected to critical loading conditions. The model in the photograph is shown being prepared for a vibration test in which the vibrations produced during launching are simulated.
(From "Flugkörper", 1960, No. 12, p. 380.)

Bild 11. Zeugnis über die Zuverlässigkeit («reliability») der Bindefunktion der Äthoxylin-(Epoxy)-Harze bei einem industriell gefertigten Verbundkörper («composite») legten die an Hand einer Gauss'schen Verteilungskurve aufgezeichneten Scherfestigkeitswerte, welche an Probestücken gemessen wurden, die einem «Vautour»-Flügel von SNCASO entnommen worden waren.
(Vergleiche auch: C. Thomas, Chief of Structures, Sud Aviation, Courbevoie, France, Vortrag «Product Bonding with ARALDITE», gehalten in Cambridge anlässlich der «Bonded Aircraft Structures» Conference, arranged by AERO RESEARCH Ltd. (heute CIBA (ARL) Ltd., 1957)

Fig. 11. Evidence of the reliability of the bonding function of the epoxies in industrially produced composites was provided by shear strength values measured on test specimens taken from a "Vautour" wing (S.N.C.A.S.O.). The Figure shows the Gaussian distribution.
(Cf. also C. Thomas, Chief of Structures, Sud-Aviation, Courbevoie, France: "Product Bonding with ARALDITE", paper read to Bonded Aicraft Structures Conference organised by Aero Research Ltd. [today CIBA (ARL) Ltd.] at Cambridge in 1957.)

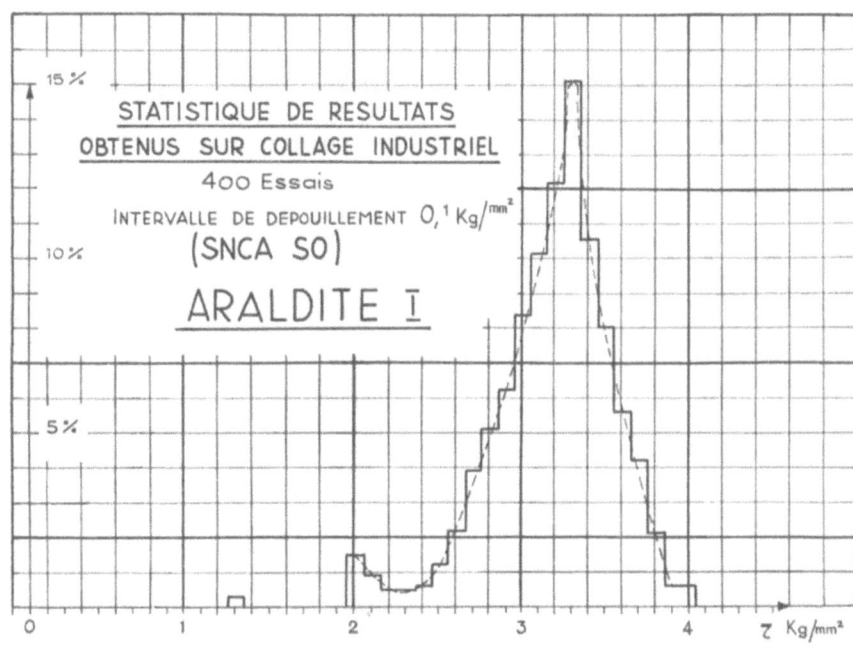

welche all die mannigfaltigen mittels Äthoxylin-(Epoxy-)Harzen hergestellten Verbundkörper («composites») des Elektro-Apparatebaues beherrscht, an Hand eines Schaubildes (siehe Bild 11) der an einem industriell hergestellten Flugzeugflügel gemessenen Scherfestigkeitswerte zur Darstellung brachte [69]. Ob irgendwelche Werkstoffkombinationen in der Starkstrom- (siehe Bild 12 und 13) oder Schwachstromtechnik/Elektronik (siehe Bild 14) mit dem neuen «Kunstharzlot hoher Isolationsgüte» [70] verbunden wurden, so zog sich durch alle Applikationen wie ein roter Faden der wichtige Qualitätsfaktor [71] der hochwertigen Verbindung unter statischen und, was in der Technik besonders bedeutungsvoll ist, dynamischen Beanspruchungen. Im angelsächsischen Sprachgebiet bildeten sich dafür die Begriffe des «Encapsulating», «Potting» («Dip-Potting»), «Packaging», «Sealing», «Impregnating» [72] u.a.m. – Bezeichnungen, die begonnen haben, auch im deutschen Sprachgebrauch sich einzubürgern. Diese Ausdrücke werden dann in der industriellen Praxis meist wechselweise gebraucht und gehen ineinander über [73]. Entscheidend und kennzeichnend ist aber, dass – wo

Bild 12. Der Stahlflansch, die Stahlschrauben und die inliegenden Glasgewebeschichten dieser 220 KV-Drucklöschkammer (MFO) sind mittels Äthoxylin(Epoxy-)Harz zu einem hochfesten Verbande vereinigt, der mit einem statischen Druck bis 220 kg/cm² belastbar ist.
(Aus: «The Role of Ethoxyline Resins in Modern Technology», 5th Int. Mech. Eng. Congress, Turin, Herbst 1953, Vortrag: E. Preiswerk, Abb. 29)

Fig. 12. The metal flange, steel screws, and the reinforcing glass cloth layers of this 220-kV quenching chamber (Oerlikon) are bonded by epoxy resins into a high-strength construction (composite) capable of withstanding a static pressure of up to 220 kg/cm².
(From "The Role of Ethoxyline Resins in Modern Technology", paper read by E. Preiswerk to 5th Int. Mech. Eng. Congress, Turin, Autumn 1953, Fig. 29.)

Bild 13. Die Kupfersegmente des Kollektors, welche durch Glimmer voneinander getrennt sind, sind mittels Äthoxylin-(Epoxy-)Harz (ARALDIT-Giessharz B) unter sich und mit dem Rohrkörper hochfest verbunden.
(Aus: «Ethoxyline Resins», E. Preiswerk, «Plastics», London, Januar/Februar 1952, Fig. 16)

Fig. 13. The copper segments of this commutator, which are insulated by mica, are bonded to one another and to the body of the rotor by epoxy resin (ARALDITE casting resin B).
(From "Ethoxyline Resins" by E. Preiswerk, "Plastics", London, January/February 1952, Fig. 16.)

immer Äthoxylin-(Epoxy-)Harze gewählt werden, um in den zwischen den zu vereinigenden Werkstoffteilen bestehenden Räumen zur Härtung gebracht zu werden – das Ganze zu einer festen, beständigen und hochbeanspruchbaren Einheit verbunden («verschweisst» [74]) wird (siehe Bild 15 und 16). Dieser Gesichtspunkt it was the quality [71] of the bond under both static and – particularly important in technical applications – dynamic (fatigue) conditions of loading that was everywhere the decisive factor. Various terms were coined for these applications: "encapsulating", "potting", "dip-potting", "packaging", "sealing", "impregnating", and numerous others [72]; in industrial practice many of these terms are used interchangeably, and to some extent overlap [73]. The all-important factor, however, is that whenever epoxy resins are used to fill the gap between two or more materials and are then cured the structure is bonded ("welded" [74]) into a strong, stable composite unit capable of withstanding extremely high stresses (see Figs. 15 and 16). This same factor has also been responsible for the success of the epoxies in the field of laminated plastics [75], and particularly glass fibre-reinforced laminates, which have also undergone rapid development during the same period (see Fig. 17) [76]. Similarly in the field of plastic tooling [77], where other plastics had already been found to give satisfactory results, the unique properties of the epoxies again proved to be decisive. Again, the use of epoxies in construction engineering (see Fig. 18) [78] and in highway engineering [79], where they perform the function of bonding or binding [80] the various component materials into a composite whole, reflects the quality of the bonding function of ARALDITE.

"It may be safely concluded that we are entering an era of composite structures-materials systems", was the accurate prediction of Dr. G. Gerard of the New York University College of Engineering in August 1959 [81]. The truth of this prediction has been amply and increasingly confirmed [82], and the epoxies are today in the forefront in every field [83] where there is a requirement for the bonding strength first communicated to the technical world in the spring of 1946 [84].

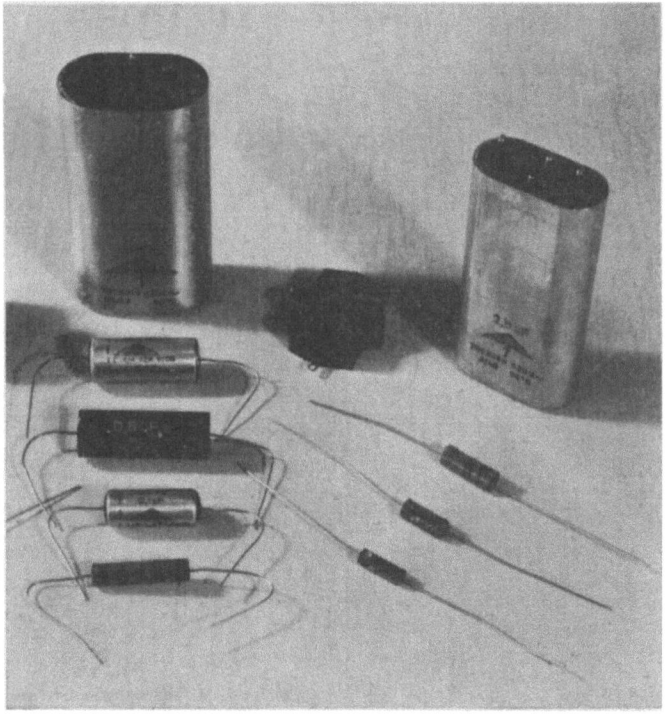

Bild 14. Von den ersten Einsätzen auf dem Schwachstrom (Elektronik)-Gebiet! «Zulöten»/«Zugiessen» von Kondensatorbechern und Teilen mit Äthoxylin-(Epoxy-)Harzen.
(Aus: «Erfahrungen über die Verarbeitung und Anwendung von ARALDIT als Bindemittel und als Giessharz», K. Meyerhans, «Kunststoffe», Dez. 1951, Abb. 14)

Fig. 14. The first applications in electronics: the "soldering" or "sealing" of capacitor bodies and other components with epoxy resins.
(From "Application and use of ARALDITE as bonding agent and as casting resin" by K. Meyerhans, "Kunststoffe", December 1951, Fig. 14.)

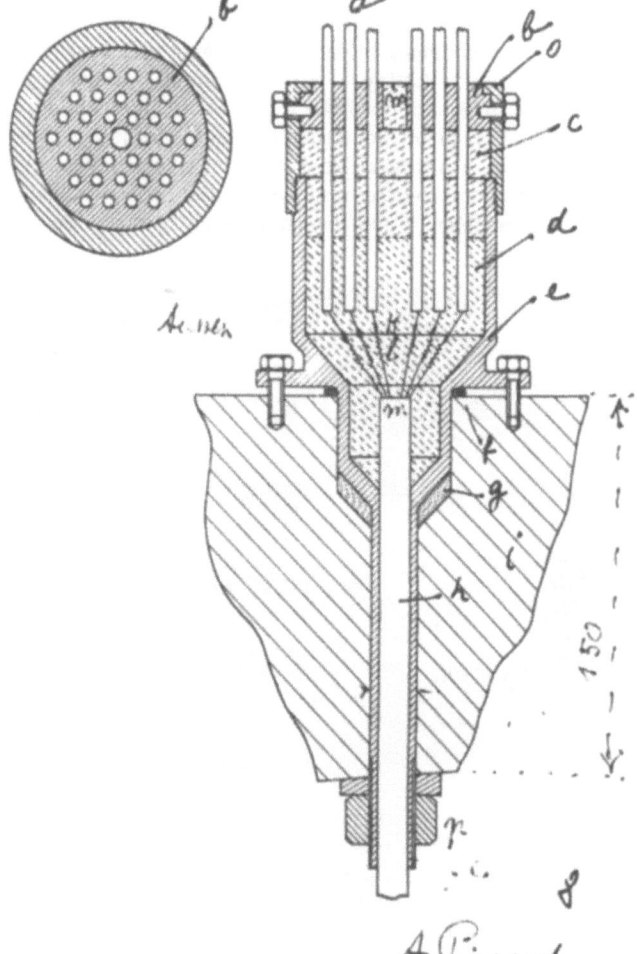

Bild 15. Die Durchführung von 36 Drähten durch die Wandung von A. Piccards Bathyscaph «Trieste» wurde mittels Äthoxylin-(Epoxy-)Harz abgesichert. Die Garantie für den zuverlässigen Verschluss gegen das Druckwasser ergab sich aus dem hochwertigen Verbande zwischen den Wandungen des eisernen Zapfens e, den Pyrotenaxkabeln a und deren Kupferseelen m, welcher durch das Äthoxylin-(Epoxy-)Harz (ARALDIT) d hergestellt wurde.
(Aus: A. Piccard, «Über den Wolken, unter den Wellen», Wiesbaden 1954, Seite 203/Fig. 22. Originalzeichnung A. Piccard)

Fig. 15. Epoxy resins were used for the protection of 36 wires passed through the hull of A. Piccard's bathyscaphe "Trieste". Complete protection against hydrostatic pressure was ensured by the high quality of the bond provided by the epoxy (ARALDITE) resin "d" between the iron plug "e", the Pyrotenax cables "a", and the copper core "m".
(From A. Piccard, "Über den Wolken, unter den Wellen", Wiesbaden 1954, p. 203, Fig. 22. Drawing by Piccard.)

Bild 16. Nicht nur die mittels Äthoxylin-(Epoxy-)Harz «verschweissten» Kupferwindungen eines Messwandlers (im Bilde links), sondern auch der mit Hilfe des gleichen Harzes «zugelötete» Tankdeckel und Tankverschluss eines Mopeds (im Bild rechts) künden im Deutschen Museum in München den Besuchern von der neuartigen Bindefunktion dieser Kunststoffe. Die beiden Schnittmodelle bilden den ersten Abteil «Äthoxylin- oder Epoxydharze» der Reihe «Neuere Kunststoffe» in der grossen Halle «Chemie» («Synthese»).(Photo Deutsches Museum München)

*Fig. 16. The novel bonding function of the epoxy resins is interestingly demonstrated by two exhibits in the Deutsches Museum in Munich: epoxy-"welded" copper windings in an instrument transformer (left) and an epoxy-"soldered" fuel tank cover and fastening for a moped (right). These two sectional models are exhibits in the section "Ethoxyline or Epoxy resins" of the display "New Plastics" in the Hall of Chemistry.
(Photograph by courtesy of Deutsches Museum, Munich.)*

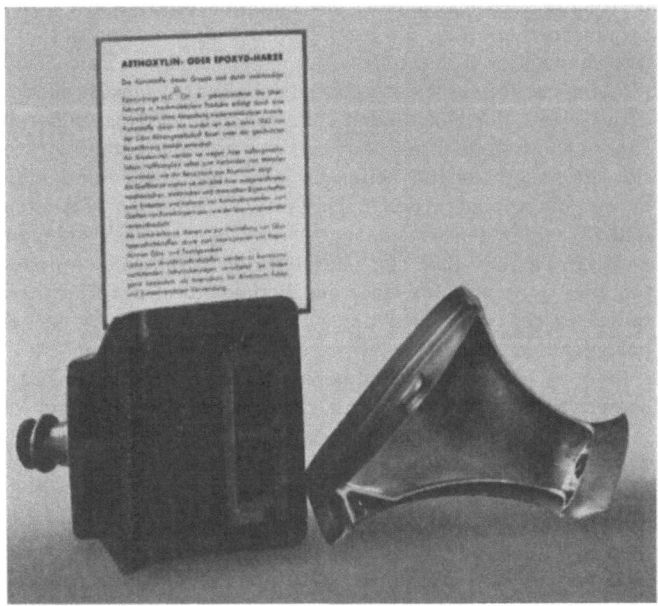

*Bild 17. Die Bindefunktion der Äthoxylin-(Epoxy-)Harze wurde auch in der sog. «Laminiertechnik» von grosser Bedeutung. Der abgebildete Apparat diente zur vergleichenden Dauerbiegewechselprüfung an Prüfstreifen, bestehend aus einer Stahllamelle, welche einseitig mit einem Prüfstreifen aus Glasgewebe/Harz-Laminat verbunden ist.
Prüfstreifen links im Bilde: Laminat, hergestellt aus Glasfasern und einem Niederdruckharz üblicher Qualität, Biegespannung 2 kg/mm², Lastspiele bis zum Bruch zwischen Metall und Harz: 1050.
Prüfstreifen rechts im Bilde: Laminat, hergestellt aus Glasfasern und Äthoxylinharz (ARALDIT). Biegespannung: 2 kg/mm². Nach 7 250 000 Lastspielen war die Verbindung Stahllamelle–Laminat noch unverändert. – Foto CIBA, Basel, Schweizer Mustermesse, 1954.
(Aus: F. K. Trietsch, «Glasgewebeschichtstoffe und ihre Verwendungsmöglichkeiten», «Konstruktion», Heft 11, 1954, Seite 434)*

*Fig. 17. The bonding function of the epoxies has also acquired great importance in laminating technology. The apparatus shown in the Figure was used for comparing the flexural fatigue strengths of specimens consisting of a steel sheet bonded on one side to a resin-glass fabric laminate.
Specimen on left of photograph: laminate composed of glass fibre and a low pressure resin of normal quality, bending stress 2 kg/mm², number of reversals of load resulting in failure of metal-laminate bond: 1050.
Specimen on right of photograph: Laminate composed of glass fibre and epoxy resin (ARALDITE), bending stress 2 kg/mm². After 7,250,000 reversals of load the steel-laminate bond was still intact. – Photograph by courtesy of CIBA Limited, Basle, Swiss Industries Fair, 1954.
(From F. K. Trietsch, "Glass fabric laminates and their application potential", "Konstruktion", No. 11, 1954, p. 434.)*

bestimmte auch massgebend den Einsatz der Harze in der sog. Laminiertechnik [75], die besonders auf dem Sektor der glasfaserverstärkten Kunststoffe während der nämlichen Zeitperiode [76] eine starke Entwicklung erfuhr (siehe Bild 17). Gerade auch auf dem Gebiete der Herstellung von Werkzeugen [77], wo bereits andere Kunststoffe ihre Brauchbarkeit erwiesen hatten, wirkten sich die neuen Gegebenheiten fördernd aus. Nicht minder forderten die applikatorischen Einsätze im Bauwesen [78] (siehe Bild 18) und in der Strassenbautechnik [79] – beim «Verbinden» und «Einbinden» [80] der verschiedenen Werkstoffkomponenten zum Ganzen – eine Güte der Bindefunktion des Kunstharzteiles, welche von Seiten der Äthoxylin-(Epoxy-)Harze in hervorragendem Masse geboten werden konnte.

«It may be safely concluded that we are entering an era of composite structures-materials systems», schrieb treffend vor 5 Jahren Dr. G. Gerard vom College of Engineering der New Yorker Universität [81]. Diese Aussage bestätigt sich in wachsendem Umfange [82] und die Äthoxylin-(Epoxy-)Harze sind die führenden Hilfsmittel [83] dort, wo jene Bindefestigkeit verlangt wird, welche durch die ARALDIT-Funktionserfindung im Herbst 1944 erkannt und im Frühjahr 1946 der Fachwelt bekanntgegeben worden ist [84].

11. Bestrebungen der internat. wissenschaftlichen Forschung zur Bestimmung der Bindefunktion der Äthoxylin-(Epoxy-)Harze

In Anbetracht der geschilderten Situation konnte nicht ausbleiben, dass in neuerer Zeit von massgebenden Vertretern der technischen Forschung, wie z.B. dem Massachusetts Institute of Technology (MIT) [85], dem Wright Air Development Center [85], dem Deutschen Kunststoffinstitut in Darmstatt [86], den Technischen Hochschulen von Hannover [87] und Karlsruhe [88], dem U.S.-Department of Commerce [89] u.a.m. [90] Ansätze zur wissenschaftlich abklärenden Durchdringung dieser Funktion unternommen wurden, die nicht allein auf der klassischen, bewährten und vielfach untersuchten Scherfestigkeitsprüfung an überlappend verklebten Metallblechstreifen beruhten [91], sondern wo man bestrebt war, die Bindefunktion am Anwendungsgegenstande (am «Verbundkörper»/«composite») selbst zu untersuchen und zu prüfen. So ergaben die Versuche der Deutschen Forschungsanstalt für Luft- und Raumfahrt e.V. (D.F.L.), Braunschweig (siehe Bild 19), um die Bindefunktion des zwischen Glasfasern gehärteten Äthoxylin-(Epoxy-)Harzes am Objekte des fertigen Laminates zu messen, bereits bemerkenswerte Resultate [92].

Es ist nicht zu bezweifeln, dass die nächsten Jahre nicht nur auf dem Sektor der tagtäglichen Anwendungen in der Praxis, sondern auch auf demjenigen der systematischen Grundlagenforschung neue Unterlagen zeitigen werden. Auch in dieser Beziehung ist das Gebiet noch voll vielversprechender Möglichkeiten. Ihre Ergebnisse stützen sich aber letztendlich auf jene Offenbarung des Jahres 1944, durch welche die seit Jahren vorliegenden Eigenschaften der Äthoxylin-(Epoxy-)Harze gedanklich derart verknüpft wurden, so dass die neue hervorstechende Eignung und die entscheidende Bedeutung dieses Kunststoff-Werkstoffes in ihrem Kerne erkannt werden konnten. Damit waren dem Harze die Wege gebahnt worden, um erfolgreich in einen weltweiten Dienst des technischen Fortschrittes zu treten.

11. The bonding function of the epoxies scientifically investigated

In view of the situation described above it was inevitable that such internationally famous centres of scientific research as the Massachusetts Institute of Technology [85], the Wright Air Development Center [85], the Deutsches Kunststoffinstitut (German Plastics Institute) of Darmstadt [86], the Technische Hochschulen (Institutes of Technology) of Hanover [87] and Karlsruhe [88], the United States Department of Commerce [89], and many others [90] should initiate systematic research programmes in order to study the bonding properties of the epoxy resins. Such research has embraced not only the classical and familiar shear strength tests performed with bonded overlapping metal strips [91] but has also included attempts to investigate the bonding function in actual finished articles (composites). Thus, interesting results have been obtained at the Deutsche Forschungsanstalt für Luft- und Raumfahrt e.V. (the German Aerospace Research Establishment) at Brunswick, where experiments were carried out (see Fig. 19) to measure the bond strength of the cured resin between the glass fibres of a laminate [92].

There can be no doubt that the next few years will bring new developments not only as regards the everyday applications of the epoxies but also in the field of systematic basic research. There is no lack of avenues waiting to be explored. In the final analysis, however, the end results of present and future work will be traceable back to the invention that was made in 1944: the process of "bisociation", whereby the familiar properties of the first ethoxyline resins were for the first time appreciated in full perspective, when the *combination* of properties was revealed and the hitherto unrevealed function of the material, with all its far-reaching implications, was recognised and identified. This discovery, the "invention of ARALDITE", made it possible for the resin to play its full part in the service of technological and scientific progress.

Bild 18. Die Firma SHELL verbindet den Lockersand (Sandkörner) in ihren Erdölbohrstellen mittels Äthoxylin-(Epoxy-)Harz zu solcher Festigkeit, wie dies bis anhin mit keinem andern organischen oder anorganischen Bindemittel gelang. Die mikroskopische Vergrösserung zeigt die Harzpartie zwischen den Sandkörnern.
(Aus: «Forschung in der Industrie»; «Die Laboratorien der Royal Dutch SHELL in Holland», «Neue Zürcher Zeitung» (Technik), Nr. 5063, 25. November 1964)

Fig. 18. The Royal Dutch Shell Company uses epoxy resins to bind loose sand grains in the drilling of oil wells; the strength obtained by this method is far greater than that hitherto possible with any other organic or inorganic binder. The photomicrograph shows the resin between the sand grains. (From "Die Laboratorien der Royal Dutch Shell in Holland", "Neue Zürcher Zeitung", Supplement on Industrial Research, No. 5063, 25 November 1964.)

Anmerkungen:

[1] «Neue Zürcher Zeitung», Zürich, Nr. 1448 (19. 4. 61), Beilage Technik, Bericht über Mustermesse Basel, Kapitel «Kunststoffe»: «...Die Entwicklung der Polyepoxyde, auch als Epon- oder Äthoxylinharze bezeichnet, nahm, was ganz besonders hervorgehoben sei, ihren Ausgangspunkt in der Schweiz, und von hier aus, oder noch genauer von Basel aus, trat dieser so interessante Kunststoff (als ARALDIT) seinen Siegeszug in die Welt an»...
Ch. F. Pitt (UNION CARBIDE) in «Modern Plastics Encyclopedia, Issue for 1959», Aufsatz «Epoxy Resins», S. 104: «...The earliest commercial materials' main use was as adhesives for metals – an outstanding development of World War II...»...
Aufsatz über SHELL's EPIKOTE-Harze in «The Rubber Age and Synthetics», Sept. 1953, S. 301: «...The earliest commercial materials, the ARALDITES, were produced by CIBA of Basle in Switzerland. Their use as adhesives was quite an outstanding development during the war»...
F. K. Trietsch (CIBA Wehr), Schrift «Die Metallverklebung», Deva Fachverlag, Stuttgart, 1960, S. 20 (Abschnitt 3.2 Epoxyharze): «...Die günstigen Verarbeitungsbedingungen, Härten ohne Druckanwendung selbst bei Raumtemperatur, ermöglichen eine universelle Anwendung der Epoxyharze... insbesondere, da ihre Klebverbindungen hohe Zugscherfestigkeiten erzielen. Ihre bekanntesten Vertreter sind die ARALDIT-Bindemittel (D.B.P. 947632, 958860, 964978)»...
«Artillerie, Armee und Technik», Nr. 2, Bern, Febr. 1964 (Zeitschrift für Artillerie, allgemeine und technische Fragen der Armee), Aufsatz: «Glas, Werkstoff der Raumfahrtindustrie», S. 9: «...In den vierziger Jahren trat eine schweizerische Entwicklung auf diesem Gebiet – die Epoxyharze, besser bekannt unter dem Namen «Araldite» – vor die Öffentlichkeit. Die Klebefähigkeit dieser Harze ist enorm, sowohl in bezug auf die erreichte Festigkeit der Klebeverbindung, als auch auf die Fähigkeit, die verschiedensten Stoffe untereinander zu verbinden. Damit waren die Voraussetzungen gegeben, um Maschinen... zu bauen, die auf ein sich drehendes Formstück Fiberglasbänder in genau vorgeschriebenem Winkel aufbringen und verkleben konnten»... ...«Die Herstellung der riesigen Gehäuse für die Weltraumrakete wäre heute ohne die Entwicklung auf diesem Gebiete völlig undenkbar.»

[2] In der breiten Öffentlichkeit – auch im schweizerischen Fernsehen – stand diese Erfindung bereits im Zentrum interessierter Betrachtung:
«National Zeitung», Basel, Nr. 440 (24. 9. 63), Abschnitt: «Der T.V.-Kommentar, Vom und über den Fortschritt»:
«...Am Sonntagnachmittag und bezeichnenderweise im familiären «Magazin der Woche» bekam man aber der Abwechslung halber Nachricht von einer Erfindung, der nichts Schreckhaftes eignet, so bahnbrechend sie auch wirkt, und die für den Schweizer besonders erfreuliche Aspekte aufweist, weil sie von einem Schweizer entdeckt worden ist und sogar in Basel verwertet wird. Es handelt sich um einen Kunstharzklebstoff... Welch unbeschränkte Möglichkeiten sich mit dieser Erfindung verbinden, zeigten Versuche im Labor... und verschiedenste Anwendungsarten... M. I.».
«technica», Basel, Okt. 1963, Rubrik «Veranstaltungen»: «...Am meisten freuen einen natürlich alle jene Anwendungen, die überhaupt erst durch den ARALDIT-Werkstoff möglich geworden sind. Hier hat denn auch der Beratungsdienst der CIBA grosse Pionierarbeit geleistet. Das Verdienst ihrer Mitarbeiter ist es doch, dass aus einem anfänglich für zahnärztliche Zwecke entwickelten Giessharz (Selitrol) eine weltweite Erfindung mit breitestem Spektrum geworden ist, die in vielen technischen Arbeitsgebieten revolutionierend wirkt. ARALDIT-Epoxydharze sind darum ein Stück Werkstoffgeschichte geworden, die den Ausbau und die Ausweitung eines Basler Grossbetriebes wesentlich bestimmt hat... H. B.».
In einem doppelseitigen Inserat («Saturday Evening Post», Philadelphia, Nov. 7, 1964, S. 44/45) nennt die Firma SHELL bei der Aufzählung ihrer Haupttätigkeitsgebiete neben den «chemicals which vastly increase crop yields» und neben den «more efficient gasolines and motor oils» als drittes und letztes die «*adhesives stronger than the materials they join*.» Sie vermerkt dies bei der Erwähnung der «invention at SHELL, or *anywhere else* in the world .»

[3] R. Müller (CIBA Wehr) in «Technische Rundschau», Bern, Nr. 17 (19. 4. 63), S. 9, Aufsatz: «Epoxydharze-Verarbeitung und Anwendung in der Industrie»: «...Im Juli 1945 erfolgte die erste Patentanmeldung auf dem Anwendungsgebiet der Verklebung (Schweiz. Patent 251647 vom 12. 7. 1945) und im Frühjahr 1946 wurde das erste Bindemittel auf Basis von Epoxydharzen während der Schweizer Mustermesse den interessierten Industrien vorgestellt.»

[4] «Kunststoffe-Plastics», «Internat. Zeitschrift für das gesamte Kunststoffgebiet», Solothurn, Nr. 3/1959, Redaktioneller Aufsatz «Die Situation der Kunststoffindustrie im Hinblick auf «Kunststoffe 59» in Düsseldorf», Kapitel «Technischer Fortschritt durch Kunstharzkleber», S. 304: «In der neueren Zeit ist die Technik des Klebens auf Anwendungsgebiete ausgedehnt worden, an die man früher nicht einmal in den kühnsten Träumen dachte... heute eine Selbstverständlichkeit... und nur wenige denken daran, dass die

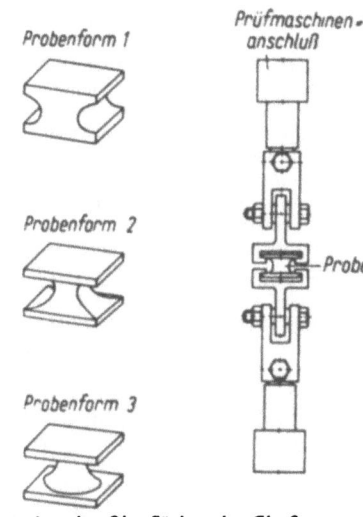

*Bild 19. Die Bindefunktion des zwischen den Oberflächen der Glasfasern gehärteten Äthoxylin-(Epoxy-)Harzes wird hier am Objekte des fertigen Laminates selbst gemessen. Die verschiedenen Probeformen wurden der Laminatplatte entnommen. Ein solches Vorgehen erfolgt als Weiterentwicklung und im Gegensatz zu den bisher bekannten Verfahren, bei welchen diese Bindefunktion unter idealisierten Verhältnissen und an eigens dazu hergestellten Proben (z. B. überlappten Blechstreifen) bestimmt wird.
(Aus: G. Niederstadt (D.F.L) in «Kunststoffe», Juni 1963, Abb. 1)*

*Fig. 19. The bonding function of the epoxy resin cured between the surfaces of the glass fibres is here shown being measured on test specimens cut from the finished laminate. The procedure illustrated is a development of, although differing from, the methods used previously in which the quality of the bonding function is measured under idealised conditions with specimens specially prepared for the purpose (e.g. shear strength tests with overlapping bonded metal strips).
(From G. Niederstadt, "Kunststoffe", June 1963, Fig. 1.)*

Notes and references:

[1] "Neue Zürcher Zeitung", Zurich, No. 1448 (19 April 1961), supplement "Technik", report on Swiss Industries Fair, Basle, section "Plastics": "The development of the polyepoxides, also known as EPON or ethoxyline resins, began, it should be emphasised, in Switzerland; this country, or rather the town of Basle, was the base from which these new plastics, under the name of ARALDITE, spread successfully throughout the world."
Ch. F. Pitt (Union Carbide) in "Modern Plastics Encyclopedia", issue for 1959, article "Epoxy Resins", p. 104: "...The earliest commercial materials' main use was as adhesives for metals – an outstanding development of World War II...."
Article on Shell's EPIKOTE resins in "The Rubber Age and Synthetics", Sept. 1953, p. 301: "...The earliest commercial materials, the ARALDITES, were produced by CIBA of Basle in Switzerland. Their use as adhesives was quite an outstanding development during the war...."
F. K. Trietsch (CIBA Wehr, Germany), in "Die Metallverklebung", Deva-Fachverlag, Stuttgart 1960, p. 20 (section 3.2, Epoxy resins): "The epoxy resins can be regarded as universal adhesives in view of their ease of application and the fact that they cure even at room temperature without pressure having to be applied...and in particular because the adhesive joint possesses good tensile shear strength characteristics. The best-known representatives of this class are the ARALDITE bonding agents (Fed. Ger. Pats. 947,632, 958,860, 964,978)."
"Artillerie, Armee und Technik", No. 2, Berne, February 1964, article entitled "Glass as a constructional material in the space industry", p. 9: "During the nineteen-forties a Swiss development in this field – the epoxy resins, better known under the name 'Araldite' – was announced. The adhesive power of these resins is outstanding, not only as regards the strength of the joint but also because they are capable of joining a vast range of different materials together. One logical consequence was the development of machines capable of winding fibreglass fabric at precisely the desired angle around a mandrel and to bond it firmly in place.... Without the progress that has been made in this field the construction of the enormous casings needed for space vehicles would have been totally inconceivable."

[2] The invention has already received a full measure of public attention and interest, not only in the Press but also on radio and television. Thus:
"National-Zeitung", Basle, No. 440 (24 September 1963), review of television programme: "Technical progress". "On Sunday

ersten bahnbrechenden Versuche erst im Zweiten Weltkrieg erfolgten... Ermöglicht wurde diese interessante Klebetechnik durch neuartige Kunststoffkleber (z. B. ARALDIT der CIBA), also durch Erzeugnisse der Kunststoffchemie»...

«Die Technologie der Klebstoffe», Teil I, von Dipl.-Ing. C. Lüttgen (Mitglied des Deutschen Patentamtes, München), 1959, S. 575: «...Die Herstellung metallischer Konstruktionen durch Verkleben der metallischen Werkstücke, noch vor gar nicht langer Zeit als Wunder bestaunt, ist nunmehr auch zu einer Selbstverständlichkeit geworden und gehört zum Allgemeinwissen des Technikers oder technisch interessierten Menschen»... (Vgl. Anm. 84).

[5] E. Preiswerk und A. von Zeerleder in «Schweizer Archiv für Angewandte Wissenschaft und Technik», Solothurn, Heft 4, 12. Jahrgang, 1946, Aufsatz: «ARALDIT, ein neues Kunstharz zum Verbinden von Leichtmetallen», Abb. 3 «Dauerbiegewechselfestigkeit (DVL) einer ARALDIT-Verbindung.» – Siehe auch Fig. 1 in Aufsatz «Plastic Bonding of Light Metals» by E. Preiswerk and A. von Zeerleder in «Plastics» Temple Press, London, Juli 1946. – Vergleiche auch z.B. die Deutsche Auslegeschrift 1089545 der Henkel & Co. GmbH, Düsseldorf (publ. 22. 9. 60), S. 1, Spalte 1: «...Vergiessen von Hohlräumen, besonders in elektrischen Apparaturen, Kabelverbindungen. In allen diesen Fällen werden die Polyepoxydverbindungen... verwendet. Nun sind für den Gebrauchswert... einer Vergussmasse neben hoher Klebekraft die Dauerhaftigkeit der Verbindung die des Materiales selbst von grosser Bedeutung. Zur Bestimmung der Dauerfestigkeit von Klebverbindungen bedient man sich neben der Erprobung in der Praxis besonders der Messung der Wechselbiegefestigkeit, wie dies z. B. von K. Meyerhans in der Zeitschrift «Kunststoffe», 41 (1951), S. 368, beschrieben worden ist»... (Diese «von K. Meyerhans beschriebene Messung der Wechselbiegefestigkeit» ist die Abschrift der Abb. 3 des obigen Aufsatzes E. Preiswerk und A. von Zeerleder.)

[6] Schon anlässlich der Mustermesse 1946 wurden dem Besucher des Standes der CIBA Muster- und Modellstücke einer Kombination von Leichtmetallblech, Glasgewebe und Äthoxylin-(Epoxy-)Harz (ARALDIT Typ I) gezeigt. Diese erlaubten Fragen wie spezifisches Gewicht/Steifigkeit/Festigkeit bei mit Hilfe von ARALDIT-Harzen hergestellten Werkstoffkombinationen zu diskutieren.

[7] Wortlaut des Vortrages abgedruckt in «Kunststoffe – Plastics», Solothurn, Nr. 4/1957, S. 151 ff. In einer Vorbemerkung bezeichnete die Redaktion diese Ausführungen als die «prägnanteste Kurzfassung der Weltgeschichte der Kunststoffe». Prof. Vieweg ist seit 1960 Präsident des Internat. Komitees für Mass und Gewicht.

[8] «Kunststoffe», Bd. 41, Dez. 1951, Aufsatz: «Wesen und Probleme der Kunststoff-Forschung», S. 407.

[9] In übereinstimmender Weise liessen sich auch die Vertreter der Fachindustrie hören. So z. B. Dr. R. Köhler (Henkel & Co. GmbH, Düsseldorf) in «Kunststoffe», Bd. 48, Okt. 1958, Aufsatz: «Zur Systematik der synthetischen Klebstoffe», S. 444: «...Die in der letzten Spalte aufgeführte Verbindung von Metallen durch Klebung erschien bis vor kurzem völlig unmöglich. Sie ist das Anwendungsgebiet der Polyadditionsprodukte, vor allem der Epoxydharze...».

[10] Eine solche Darstellung war durchaus am Platze, denn auch z.B. im Deutschen «Aluminium-Taschenbuch» des Jahres 1942 wird im Abschnitt «Kleben und Kitten» ausgeführt (S. 285): ...«Ansprüche an Festigkeit können bei Kittverbindungen nicht gestellt werden»... In der 11. Auflage (1955) des gleichen Taschenbuches (Aluminium-Zentrale e.V., Düsseldorf) wird erneut treffend bestätigt (S. 478): ...«Bis vor gut 10 Jahren dienten Klebverfahren, abgesehen von grossflächigen Verbindungen wie Folien-Kaschieren, Metall-Holzfurnieren oder andern Belägen von Aluminiumblechen, im allgemeinen nur für untergeordnete Zwecke. Besondere Ansprüche an Festigkeit konnten derartige Verbindungen nicht erfüllen. Metalle wurden deshalb nur gelegentlich verklebt. Die Entwicklung der letzten Jahre führte jedoch zu Klebstoffen, die Verbindungen auch bei hochbeanspruchten Konstruktionen zulassen. Sie ist heute in vollem Fluss»...

[11] «Light Metals», London, Juli 1940, S. 175/176.

[12] «...The possibility of partially or wholly displacing the traditional methods cited, by newer processes involving the uses of adhesives, is not envisaged; rather it is anticipated that an expansion of the recognized standard (soldering, brazing, welding, riveting) means for joining metal to metal or to other substances will follow»...

[13] «...Little or no published information is available upon such compounds and upon their appropriate fields of use, whilst in general their composition, and the changes they undergo in setting or hardening, are so complex as to make it impossible to deduce their suitability, or otherwise, from straight-forward considerations»...

[14] In der wöchentlich erscheinenden Zeitung «Technische Rundschau und Allgemeine Industrie und Handelszeitung», Bern, vom 28. 1. 44, wurden in einem kurzen Artikel «Kleben von Metallen» Vermutun-

afternoon, and, significantly, during the family programme 'Weekly Magazine', we had the unusual opportunity of learning about an invention that represents a genuine advance in modern technology. It is of particular interest to Swiss viewers, since it was invented by a Swiss and is under development here in Basle. It is a plastic adhesive.... Its unlimited potential was impressively documented by films of laboratory tests and of many of the actual applications in which it is used.... M.I."

"technica", Basle, October 1963: "The most gratifying applications are naturally those that but for ARALDITE would never have been possible. This is where CIBA's advisory service has done pioneering work; for it was by the efforts of this department that SELITROL, a casting resin originally developed for dental applications, became the basis for an invention that has become of worldwide importance and that is causing a revolution in numerous areas of technology. ARALDITE epoxy resins thus represent an important chapter in the history of materials technology and have been to a great extent responsible for the expansion and growth of a large Basle company.... H.B."

In a two-page advertisement (in the "Saturday Evening Post", Philadelphia, 7 November 1964, pp. 44 and 45), the firm of Shell listed its principal fields of activity as "chemicals which vastly increase crop yields", "more efficient gasolines and motor oils", and, in third and last place, "adhesives stronger than the materials they join", mentioning their "invention at Shell, or anywhere else in the world".

[3] R. Müller (CIBA Wehr, Germany), in "Technische Rundschau", Berne, No. 17 (19 April 1963), p. 9, article entitled "Application and use of epoxy resins in industry": "In July 1945 the first patent application was made in respect of the application of the resin as an adhesive (Swiss Patent No. 251,647, dated 12 July 1945), and in the spring of 1946 the first epoxy resin-based bonding agent was introduced to representatives of industrial companies at the Swiss Industries Fair."

[4] "Kunststoffe-Plastics", Solothurn, No. 3, 1959, editorial under the title "The present position of the plastics industry as reflected by the 'Kunststoffe 59' exhibition in Düsseldorf", in the section headed "Technological progress due to plastics adhesives", p. 304: "In recent years adhesives technology has been extended to cover new fields of application which a short time ago even the most optimistic person would have never dreamt of but which are now taken for granted, and it is hard to remember that the first pioneering experiments were undertaken as recently as during World War II.... This interesting development was made possible by 'plastics adhesives' (e.g. CIBA's ARALDITE), i.e. by products of the plastics industry."

"Die Technologie der Klebstoffe", Part I, by C. Lüttgen (member of German Patent Office, Munich), 1959, p. 575: "...The construction of metallic structures by the bonding together of metallic components was until quite recently regarded as impossible but has now become commonplace, part of the normal fund of knowledge of every technologist or indeed of everyone interested in technical matters...." See also Note [84] below.

[5] E. Preiswerk and A. von Zeerleder, in "Schweizer Archiv für Angewandte Wissenschaft und Technik", Solothurn, No. 4, Vol. 12, 1946, article entitled "ARALDITE, a new synthetic resin for bonding light alloys": Fig. 3 "repeated flexural fatigue strength of an ARALDITE bonded joint". – See also Fig. 1 in the article "Plastic Bonding of Light Metals" by E. Preiswerk and A. von Zeerleder in "Plastics", Temple Press, London, July 1946.

See also, for example, the German patent specification No. 1,089,545 (Henkel & Co. GmbH, Düsseldorf, published 22 September 1960), p. 1, col. 1: "The filling of hollow spaces, particularly in electrical apparatus and cable joints: the polyepoxide compounds are used for all these applications. The practical value of a casting resin is dependent not only on its bonding power but also on the durability of the bond. The endurance of an adhesive joint is tested not only with the aid of finished articles but also, in particular, by means of the repeated flexural strength test as described, e.g., by K. Meyerhans in the journal 'Kunststoffe', No. 41, 1951, p. 368." – The "repeated flexural strength test as described by K. Meyerhans" is a reproduction of Fig. 3 of the article by Preiswerk and von Zeerleder referred to above.

[6] At this same Swiss Industries Fair in 1946 visitors to the CIBA stand were also shown specimens of a combination of light alloy sheet, glass fabric, and epoxy resin (ARALDITE Type 1). Such questions as the specific gravity, rigidity, and strength of composites produced with the aid of ARALDITE resins could thus be considered in some detail.

[7] Reprinted (in the original German) in "Kunststoffe-Plastics", Solothurn, No. 4 (1957), pp. 151 ff. In an editorial introduction Professor Vieweg's remarks are described as the "most valuable short summary of the history of plastics". Professor Vieweg has been President of the International Committee for Weights and Measures since 1960.

[8] "Kunststoffe", Vol. 41, December 1951, article entitled "Nature and problems of plastics research", p. 407.

gen über die Natur der «geheimen» amerikanischen und englischen Bindemittel angestellt, auf welche in der Tagespresse und in technischen Zeitschriften hingewiesen wurde. Der Sachverständige erkannte übrigens sofort, dass die vom Autor aufgezählten Harze die angestrebte Lösung nicht bieten konnten.

[15] Vergleiche z. B. den «Redux»-Prozess im Aufsatze «The Gluing of Light Metals» in Zeitschrift «Light Metals», Mai 1947, S. 234 ff.

[16] «The Technology of Adhesives», by John Delmonte, Reinhold Publ. Corp., New York, 1947, Chapter 17: «Adhesives for Metal and Rubber», S. 424 ff.

[17] Vergleiche die «Entwicklungsstory» von Devoe & Raynolds, dargestellt in der US-Zeitschrift «Chemical Processing», Aug. 1955 S. 10 ff., unter dem Titel «Pioneers of Progress», «Epoxy Resins – product of deliberate research.» Mit dem ersten, allerdings erst am 14. 9. 43 angemeldeten US-Patent 2456408 eröffnete S.O. Greenlee eine lange Reihe von Patenten über synthetische Arbeiten auf dem Äthoxylin-(Epoxy-)Harzgebiet, welche aber praktisch nur auf die Entwicklung von Lackharzen ausgerichtet waren. Vergleiche dazu auch C. E. Hutz (selig) in «Manufacure of Plastics», Vol. I, Reinhold, New York, 1964, S. 498–500. Siehe auch SHELL's erste Broschüre «EPON Surface Coating Resins», SHELL Chem. Corp., San Francisco/New York, Copyright 1948, SC: 48–9. Auf jenes erstmalige Erscheinen auf dem Markt weist auch J. Kistler im Aufsatz «Ein neues SHELL-Kunststofflaboratorium» in «Chem. Rundschau», Nr. 14, 15. Juli 1960, S. 366, hin.

[18] DRP 676117 (angemeldet 11. 12. 34, ausgelegt 5. 1. 39, druckschriftlich ausgegeben 26. 5. 39). Vergleiche auch A.M. Paquin, «Epoxydverbindungen und Epoxydharze», Springer-Verlag, Berlin, 1958, S. 310/311.

[19] Vergleiche Dr. W. Fisch in Houwink-Staverman, «Chemie und Technologie der Kunststoffe», II/2, 4. Auflage, Leipzig 1963, S. 1015.

[20] Vergleiche A.M. Paquin, loc. cit. Anm. 18, Vorwort Seite XI mit Anmerkung 1.

[21] Betreffend des Begriffes «Formulieren» vergleiche Abschnitt 8 und Anm. 47.

[22] Über jenen Stand der Technik vergleiche: Houwink «Chemie und Technologie der Kunststoffe», Band II, Leipzig 1942, S. 185, 186 und 187.

[23] Vergleiche Ullmann «Encyclopädie der Technischen Chemie», Bd. 5, 1954, S. 710 ff (Dentalchemie): «...Dem PALADON entsprechende Werkstoffe... heute das bevorzugte Material der Zahnprothetik... ausreichende mechanische Festigkeit... leichte Reparaturmöglichkeit»...

[24] Auch in der Arbeit E. Dolder «Physikalische Werkstoffprüfungen an Zahnprothesen im Laboratorium und am Patienten», «Schweiz. Monatsschrift für Zahnheilkunde», Band 53, 1943, wurde die durch die ARALDIT-Erfindung geschaffene Erkenntnis nirgends offenbart.

[25] Vergleiche Handbuch «Engineering Properties and Applications of Plastics», G. F. Kinney, New York, 1957, Kapitel: Plastics/Polymethyl Methacrylate, S. 54/55: «These dentures are quite successful and, substantially, all synthetic dentures made in the United States are of acrylic plastic»...

[26] Vergleiche: G. F. Kinney (siehe oben loc. cit. Anm. 25), S. 134 (Kapitel: «Epoxies»): «It can be observed that the adhesive nature of the epoxies is something of an advantage in castings, as used in the potting (encapsulating) of electric components but, for some purposes, this may be troublesome. Specialty mold release agents have been developed for use in the production of cast epoxy drawing dies, laminates, and moldings.»
F.C. Hopper der SHELL Development Company in dem unter USAF-Contract Nr. AF 33(038)-19587 für die USAF, Wright-Patterson Air Force Base, Ohio, angefertigten Report 52-5, Suppl. 3 «High Strength EPON Laminates», S. 57, 3. Absatz: ...«A prominent characteristic of EPON Resins is their strong adhesion to most surfaces against which they may be cured. It is this property which accounts, at least in part, for the high strength obtainable in EPON resin laminates. This property is also the basis for the problem of satisfactory release from molds and cauls.»

[27] Vergleiche z. B. «Gebrauchsanweisung; de Treys Prothesenmaterial SELITROL», S. 3, Herausgeber: Gebr. de Trey AG, Zürich.

[28] Direktor der Firma Moser-Glaser & Co. AG, Muttenz bei Basel, Autor des bekannten Handbuches «Elektrische Isolierstoffe», Zürich, 1. Auflage 1946.

[29] Dipl.-Ing. A. Imhof war damals am Ausstellungsstande der CIBA von Dr. Preiswerk empfangen und von diesem über die neuen Äthoxylinharze orientiert worden.

[30] Vergleiche z. B.:
A. Imhof: «Fortschritte im Bau von Trockentransformatoren und Messwandlern», «Schweiz. Techn. Zeitschrift», Bd. 44, 1947, S. 760 ff.
«Diskussionsbeiträge/SVMT-Kunststofftagung 31. 1. 1948», «Schweiz. Archiv für Angewandte Wissenschaft und Technik», Bd. 15 (1949), S. 286.
«Einige Problemstellungen der Elektrotechnik an die Kunststoffchemie», «Schweiz. Techn. Zeitschrift», Bd. 46, 1949, S. 626 ff.

[9] Representatives of the plastics and metals industries also expressed their full agreement with this opinion. Thus R. Köhler (Henkel & Co. GmbH, Düsseldorf), in "Kunststoffe", Vol. 48, October 1958, article entitled "Systematics of synthetic adhesives", p. 444: "The bonding of metals by adhesives described in the last column had until recently been regarded as totally impossible; this function is, however, now performed by polyaddition products, and in particular by the epoxy resins."

[10] This point of view was at the time entirely justified; similarly in the "Aluminium-Taschenbuch", published in Germany in 1942, it is stated in the section on "Glues and Cements" (p. 285): "It is not reasonable to expect cemented joints to exhibit any significant degree of strength." In the 11th edition of this manual (Aluminium-Zentrale e.V., Düsseldorf), however, we read: "Until some ten years ago processes using adhesives, other than those involving large surfaces such as film-coating, metal-facing of wood, and other types of application where aluminium sheets were used for coating, were of purely secondary importance. The resulting combinations of materials were unable to satisfy special requirements where bond strength was necessary, and as a result adhesives were only rarely used for bonding metals. Developments of the last few years have, however, resulted in adhesives that enable metals to be bonded even under severe conditions: development still continues apace."

[11] "Light Metals", London, July 1940, pp. 175 and 176.

[12] "...The possibility of partially or wholly displacing the traditional methods cited by newer processes involving the use of adhesives is not envisaged; rather it is anticipated that an expansion of the recognised standard (soldering, brazing, welding, riveting) means for joining metal to metal or to other substances will follow...."

[13] "...Little or no published information is available upon such compounds and upon their appropriate fields of use, whilst in general their composition, and the changes they undergo in setting or hardening, are so complex as to make it impossible to deduce their suitability, or otherwise, from straightforward considerations...."

[14] The weekly publication "Technische Rundschau und Allgemeine Industrie- und Handelszeitung", Berne, which was obtainable at practically any news-stand, contained a short article on 28 January 1944 entitled "Gluing of Metals" speculating on the nature of the "secret" British and American bonding agents that had been mentioned in the daily press and in technical trade journals. A specialist was, however, able to recognise immediately that the resins mentioned by the author, e.g. pure phenolics, ureas, and melamines, would not be capable of providing the desired solution which was subsequently provided by the epoxies.

[15] See, for example, the description of the "Redux" process in the article "The Gluing of Light Metals" in "Light Metals", May 1947, pp. 234 ff.

[16] "The Technology of Adhesives", by John Delmonte, Reinhold Publ. Corp., New York 1947, chap. 17: "Adhesives for Metal and Rubber", pp. 424 ff.

[17] See the article on Devoe & Raynolds in the U.S. journal "Chemical Processing", August 1955, pp. 10 ff., under the title "Pioneers of Progress": "Epoxy resins – product of deliberate research". With the first American patent (No. 2,456,408, which however was applied for only on 14 September 1943), S. O. Greenlee began a long series of patents covering the synthesis of epoxy resins; this work, however, was directed almost exclusively to the development of resins for paints and varnishes.
See also the comments of the late C. E. Hutz printed in "Manufacture of Plastics", Vol. I, Reinhold, New York 1964, pp. 498–500, and Shell's first brochure on "EPON Surface Coating Resins", published by Shell Chem. Corp., San Francisco and New York, copyright 1948, SC: 48-9. The first appearance of these resins on the market is also referred to by J. Kistler in his article entitled "A new Shell plastics laboratory", "Chemische Rundschau", No. 14, 15 July 1960, p. 366.

[18] German Pat. 676,117 (applied for on 11 December 1934, made available for inspection on 5 January 1939, published on 26 May 1939). See also A. M. Paquin, "Epoxydverbindungen und Epoxydharze", Springer-Verlag, Berlin 1958, pp. 310 and 311.

[19] See W. Fisch in Houwink-Staverman, "Chemie und Technologie der Kunststoffe", II/2, 4th Ed., Leipzig 1963, p. 1015.

[20] See A. M. Paquin, loc. cit. (Note [18] above), introduction, p. xi, and footnote 1.

[21] See section 8 and also Note [47], regarding the meaning of the word "formulation".

[22] Regarding the state of knowledge at that time see Houwink, "Chemie und Technologie der Kunststoffe", Vol. II, Leipzig 1942, pp. 185–187.

[23] See Ullmann, "Encyclopädie der Technischen Chemie", Vol. 5, 1954, pp. 710 ff. (dental chemistry): "...materials of the type PALADON, currently the material of choice in prosthetic dentistry...adequate mechanical strength...ease of repair".

[24] The paper by E. Dolder "Physical testing of materials for dentures in the laboratory and in the patient", in "Schweizerische

«Ein neuer Trocken-Spannungswandler», Bulletin SEV 1949, Heft 13.
Siehe auch den Aufsatz «Ethoxylines, a new group of triple-function resins» (Dr. E. Preiswerk and Dr. C. Meyerhans) in «Electrical Manufacturing» New York, Juli 1949, wo auf S. 78 eine Photoabbildung eines kleinen Messwandlers von Moser-Glaser und auf S. 166 die Erwähnung dieser Firma zu finden ist. Diese Veröffentlichung beinhaltete im wesentlichen die Ausführungen eines Vortrages von Dr. Preiswerk am 31. 1. 48 in Zürich anlässlich der Kunststoff-Tagung des Schweiz. Verbandes für Materialprüfung in der Technik (SVMT). Dieser Vortrag ist erwähnt als Anmerkung 3 auf S. 24 des «Schweiz. Archiv für Angewandte Wissenschaft und Technik», Januar 1949.

[31] Vergleiche z.B.: H. Koller, «Neue Trockenstromwandler mit Kunstharzisolation», «Bulletin SEV», 1950, Heft 1.
A. Gantenbein, «Neue Giessharzisolationen in der Hochspannungstechnik», «Scientia Electrica», Heft 2, 1954.

[32] Vergleiche A. M. Paquin, loc. cit., Anm. 18, S. 309: «...In einer nicht allzu langen Vergangenheit hatte man kein Interesse für Harze, und man konnte sich nicht vorstellen, dass sie irgendwelche technische Aufgaben erfüllen könnten. Bei ihrer unbeabsichtigten Bildung wurden sie missachtet, ihre Eigenschaften wurden nicht studiert, und ihrer in der Literatur keine Erwähnung getan»...

[33] Vergleiche A. M. Paquin, loc. cit., Anm. 18, S. 631: «...Beim Vergiessen von Harzen liegt allgemein das schwerwiegende Problem der Loslösung des Gießstückes aus der Form vor»... (Vergleiche auch oben, Anm. 26).

[34] Akademische Verlagsanstalt Becker und Erler Kom.-Ges., Leipzig, 1940. (Vergleiche: Meyer, K. H., Natural and synthetic high polymers, 1st Ed. (1942), 2nd Ed. (1950), Vol. IV of High Polymers, Wiley, New York.)

[35] Vergleiche dazu zum Beispiel die Definition im Aufsatze «Industrial Adhesives as Market for Synthetic Resins» (P. W. Sherwood in «Chemische Rundschau», 1. 10. 63), Abschnitt: «Role of Adhesives»: «Several objectives are served by the use of adhesives, among them:...Bonding small particles into large pieces»...

[36] Vergleiche Abschnitt 6 und Anm. 34.

[37] Über die Bedeutung der Notwendigkeit einer grundlegenden, vielseitigen Ausbildung des schweizerischen Chemikers für die Wahrnehmung der wesentlichen Chancen durch die schweiz. Industrie in «ihrem zähen Ringen um den technischen Fortschritt» vergleiche: Heft 90 der «Kultur und Staatswissenschaftlichen Schriften der Eidgenössischen Technischen Hochschule», «Vom Ursprung des Technischen Fortschrittes». Der Autor (Prof. Brandenberger) hält u. a. auch die Bedeutung fest (S. 20/21/22), welche das frühzeitige «Zusammenhänge-Erkennen», «geschickte Kombinieren» und «intuitive Erfassen» – als «individuelle Begabungen» – darstellen, um «vor allem neuen Werkstoffen den Weg zu bahnen und dem technischen Fortschritt zu dienen». – Vergleiche den gleichen Autor in «Allgem. Schweiz. Militärzeitschrift», Okt. 1959, Aufsatz: «Technik und Wissenschaft im Dienste der Landesverteidigung»: «...wer auf zwei Gebieten zu Hause... vermag nicht selten den entscheidenden Impuls zu geben»...
Anlässlich der Deutschen Kunststofftagung im Okt. 1954 in Stuttgart führte Dr. rer. nat. Rudolf Gäth, Leiter der Anwendungstechnischen Abteilung für Kunststoff-Rohstoffe der Badischen Anilin und Sodafabrik (BASF) in seinem Plenarvortrage «Wo stehen wir in der Entwicklung der Kunststoffe?» u. a. aus: «...dass das gemeinsame Wissen um Metalle und Kunstharze äusserst wertvoll sei und die Entwicklung fördere...»

[38] Dem Forschungsinstitut Neuhausen der AIAG stand u. a. die Aufgabe zu, die unzähligen Produkte zu überprüfen, welche die allgemeine Technik (z. B. auch die chemische Technik mit ihren Farbstoffen, Lacken usw.) auf den Markt brachte, damit dieselben auch auf dem Gebiete der Leichtmetalle gebraucht werden könnten. Bei dieser Arbeit, die vornehmlich im Dienste der Kundenberatung stand, galt es auch ungeeignete, die Applikationen der Leichtmetalle diskriminierende Produkte und Methoden festzustellen. Es ist begreiflich, dass – beim vorliegenden Stande der Dinge – in Neuhausen bis anhin noch nie ein kraftschlüssiges Metallbindemittel auf Kunstharzbasis geprüft worden war.

[39] Betreffend der Wichtigkeit der Dauerfestigkeiten vergleiche Dr. R. Gäth, Kunststoff-Rohstoff-Laboratorium der Anwendungstechnischen Abteilung der Badischen Anilin und Sodafabrik AG (BASF), Ludwigshafen (Rhein), im Aufsatze «Grundlegende Überlegungen bei der Auswahl eines Kunststoffes». Kapitel: «Bedeutung der Dauerwiderstandsfestigkeit», «Kunststoffe», München, Okt. 1963 (nach einem Vortrage, gehalten am 19. 4. 63 auf dem Kunststoff-Kongress in Wien).

[40] Vergleiche dazu: Zeitschrift «Chemical Processing», Aug. 1955, S. 6/7: Address delivered on the occasion of the presentation to Mr. Williams (Vice-Pres. and Adviser on Research and Development, E. I. Du Pont de Nemours & Comp.) of the Perkin Medal of the American Section of the Soc. of Chem. Ind.: «Research from a management point of view. For top return on research dollars

Monatsschrift für Zahnheilkunde", Vol. 53, 1943, likewise makes no reference to the bonding function later recognised by the "invention of ARALDITE".

[25] See the manual "Engineering Properties and Applications of Plastics", G. F. Kinney, New York 1957, section on "Plastics – Polymethyl Methacrylate", pp. 54 and 55: "These dentures are quite successful and, substantially, all synthetic dentures made in the United States are of acrylic plastics."

[26] See also: G. F. Kinney (loc. cit., see Note [25] above), p. 134 (in the section headed "Epoxies"): "It can be observed that the adhesive nature of the epoxies is something of an advantage in castings, as used in the potting (encapsulating) of electric components, but for some purposes this may be troublesome. Specialty mold release agents have been developed for use in the production of cast epoxy drawing dies, laminates, and moldings."
F. C. Hopper (Shell Development Company) in a Report (No. 52–5, Suppl. 3) "High Strength EPON Laminates", prepared for the U.S.A.F., Wright-Patterson Air Force Base, Ohio, under U.S.A.F. Contract No. AF 33(038)-19587, p. 57, 3rd parag.: "...A prominent characteristic of EPON resins is their strong adhesion to most surfaces against which they may be cured. It is this property which accounts, at least in part, for the high strength obtainable in EPON resin laminates. This property is also the basis for the problem of satisfactory release from molds and cauls."

[27] See, for example, "Instructions for using de Trey SELITROL material for dentures", p. 3, published by Gebr. de Trey AG, Zurich.

[28] Manager of the company Moser-Glaser & Co. AG of Muttenz, near Basle, Switzerland, and author of the well-known compendium "Elektrische Isolierstoffe", Zurich, 1st Ed. 1946.

[29] A. Imhof was received at the CIBA stand by Dr. Preiswerk, who provided him with information regarding the new ethoxyline (epoxy) resins.

[30] See, for example:
A. Imhof: "Recent advances in the construction of dry type transformers and instrument transformers", "Schweizerische Technische Zeitschrift", Vol. 44, 1947, pp. 760 ff.
"Diskussionsbeiträge/SVMT-Kunststofftagung" held on 31 January 1948, "Schweizer Archiv für Angewandte Wissenschaft und Technik", Vol. 15 (1949), p. 286.
"Problems of electrical engineering facing plastics chemistry", "Schweizerische Technische Zeitschrift", Vol. 46, 1949, pp. 626 ff.
"A new dry type potential transformer", "SEV-Bulletin" 1949, No. 13.
See also the article "Ethoxylines, a new group of triple-function resins" by E. Preiswerk and C. Meyerhans in "Electrical Manufacturing", New York, July 1949: on p. 78 is a photograph of a small instrument transformer made by Moser-Glaser & Co. AG, and the firm is mentioned by name on p. 166. This article contains the essentials of the paper presented by E. Preiswerk on 31 January 1948 to the Plastics Congress of the SVMT held in Zurich. The paper is also referred to in Note 3 on p. 24 of the "Schweizer Archiv für Angewandte Wissenschaft und Technik", January 1949.

[31] See, for example:
H. Koller, "A new dry type current transformer with synthetic resin insulation", "SEV-Bulletin", 1950, No. 1.
A. Gantenbein, "New casting resins as insulating materials in high tension engineering", "Scientia Electrica", No. 2, 1954.

[32] Compare A. M. Paquin, loc. cit. (see Note [18] above), p. 309: "In the relatively recent past resins aroused no interest, and they were not thought of as being able to satisfy technical requirements of any kind. When they were accidentally produced they were ignored, their properties remained unstudied, and no mention was made of them in the literature."

[33] See A. M. Paquin, loc. cit. (see Note [18] above), p. 631: "Whenever resins are cast, the problem arises of releasing the cast part from the mould." See also Note [26] above.

[34] Akademische Verlagsanstalt Becker und Erler Kom.-Ges., Leipzig 1940. (Compare: K. H. Meyer, "Natural and Synthetic High Polymers", 1st Ed. [1942], 2nd Ed. [1950], Vol. IV of "High Polymers", Wiley, New York.)

[35] Compare, for example, the definition given in the article "Industrial Adhesives as Market for Synthetic Resins" (P. W. Sherwood in "Chemische Rundschau", 1 October 1963) under the heading "Role of Adhesives": "Several objectives are served by the use of adhesives, among them...bonding small particles into large pieces...."

[36] Compare section 6 and Note [34].

[37] Professor Brandenberger, writing in issue No. 90 of the "Kultur- und Staatswissenschaftliche Schriften der Eidgenössischen Technischen Hochschule", emphasises that the education of Swiss chemists should be as thorough and as varied as possible so that Swiss industry, in its "struggle" for technological progress, may at all times be in a position to make the most of every opportunity that presents itself. In his article, entitled "The origins of technical progress", he states on pages 20 to 22 that the ability to correlate facts, skilled associative thinking, and intuitive comprehension

management must: Guard against premature selling of the lab's new products. Maintain individual initiative along with smooth relationships within groups... Discovery and invention are highly individual matters, whereas the great mass of technical effort is concerned with the elaboration of ideas. We all know that a single concept by an individual can change the whole course of an industry. The research worker has a potential that is, without exaggeration, extraordinary.»

[41] Vergleiche PD Dr. Mario Pedrazzini, Mitarbeiter des als umfassendes Standardbuch geltenden Kommentars zum Schweizerischen Patentgesetz in seinem Vortrage «Starrheit und Elastizität im gewerblichen Rechtsschutz» vor dem Zürcherischen Juristenverein («Neue Zürcher Zeitung», Nr. 4470, 14. 11. 62): ...«Die Praxis ist dazu übergegangen, neue Figuren der Erfindung anzuerkennen, nämlich die sog. Funktionserfindung»...

[42] Da Holland das erste Patentland war, in welchem die Erfindung einer rigorosen Prüfung unterzogen wurde, so kam der Erteilung des Patentes besondere Bedeutung zu. Die Verhandlungen vor dem Beschwerdesenat in Den Haag fanden am 17. 10. 49 statt. Dr. Preiswerk hatte die CIBA Basel zu vertreten.

[43] Vergleiche «Sheet Metal Industries», September 1949 (London) wo der Bericht der EMPA (Prof. Dr. M. Ros, Sachbearbeiter: Dipl.-Arch. ETH H. Kühne) unter dem Titel «A new Bonding Resin, Detailed Results of Laboratory Tests on ARALDITE» zu Handen der englisch sprechenden Fachwelt veröffentlicht wurde. Anlässlich einer Besprechung des in Köln erschienenen Buches von E. Krekeler «Das Verbinden von Metallen durch Kunstharzkleber» (Forschungsbericht Nr. 246 des Wirtschafts- und Verkehrsministeriums Nordrhein-Westfalen), 1956, musste H. Kühne, EMPA, Zürich (vergleiche «Schweizer Archiv für Angewandte Wissenschaft und Technik», März, 1958, S. 94) feststellen, dass «man bei der Durchsicht des Schrifttumverzeichnisses die ausgedehnten, mit fast identischer Zielsetzung in der Schweiz ausgeführten Untersuchungen (so unter anderem «Sheet Metal Industries», Sept. 1949) vermisst». Offensichtlich waren diese Veröffentlichungen im deutschen Sprachgebiet nicht beachtet worden. – Anknüpfend an Untersuchungen über den sog. «joint factor» («Gestaltsfaktor»), welche N.A. de Bruyne (siehe z.B. «The strength of glued joints», «Aircraft Engineering», Vol. XVI, 115 (1944) bekanntgegeben hatte, konnte K. Frey an Hand der durch die Untersuchungen der EMPA angehäuften Messwerte auf rechnerischem Wege noch einige weitere gesetzmässige Zusammenhänge an einfach überlappten Blechverbindungen feststellen («Schweizer Archiv», Heft 2, 19. Jahrg., 1953, Aufsatz: «Beiträge zur Frage der Bruchfestigkeit kunstharzverklebter Metallverbindungen»).

[44] In dem wichtigen Industriegebiete der USA hatte Dr. Preiswerk Gelegenheit, Vertretern der massgebenden Technik (speziell auch Elektrotechnik, Air Material Command/Dayton u.a.m.) anlässlich eines Besuches im Herbst 1948 über die ersten Erfahrungen in Europa zu berichten. Es folgten unmittelbar darauf im Jahre 1949 die ersten Publikationen: a) «Ethoxylines, a New Group of Triple Function Resins», «Electrical Manufacturing», Juli 1949 (Preiswerk und Meyerhans); b) «New Resins provide Practical Bonding Agent for Metals», «Materials and Methods», Okt. 1949 (Preiswerk, Meyerhans und Denz).

[45] Von weiteren Mitteilungen über die nun folgende Entwicklung seien u.a. erwähnt:
E. Preiswerk, Vortrag «Ethoxyline Resins, Observations on Development up to the Present – Aspects for Future Applications» (Sept. 1951, Summer-School der AERO-RESEARCH Ltd. (heute CIBA (ARL) Ltd. in Cambridge) publiziert in: Buch «Structural Adhesives», London (Lange Ltd.), Dez. 1951; «Light Metals», London, Nov. 1951; «Plastics», London, Jan./Febr. 1952.
Der Vortrag wurde im ähnlichen Rahmen in französischer Sprache im Mai 1952 anlässlich der las Jornadas Nacionales de Plásticos in Madrid gehalten. Er wurde in der span. Kunststoffzeitschrift «Revista de Plásticos», Nr. 16, Juli/August 1952, unter dem Titel «Nuevos Desarrollos en el campo de las Resinas Etoxilínicas» publiziert.
K. Meyerhans, «Bindemittel und Giessharze auf ARALDIT-Basis», «Kunststoffe», München, Nov. 1951.
«Erfahrungen über die Verarbeitung und Anwendung von ARALDIT als Bindemittel und als Giessharz», «Kunststoffe», München, Dez. 1951 (nach einem Vortrag an der Kunststoff-Tagung in Wiesbaden, Okt. 1951).
E. Preiswerk, «Äthoxylinharze in der Elektrotechnik», Vortrag gehalten am 17. 10. 52 an der Fachtagung des VDE «Kunststoffe in der Elektrotechnik» anlässlich der Deutschen Kunststoffmesse 1952 in Düsseldorf, publiziert in ETZ, 5., Ausgabe B, Heft 1 (21. 1. 53). Auch als erweiterter Sonderdruck ausgegeben. (Im ähnlichen Rahmen erfolgte am 16. 10. 52 ein Vortrag «Gedanken zum Einsatz der Äthoxylinharze in der modernen Technik» vor dem VDI, Sektion Frankfurt a.M.).
«Le Rôle des Résines Ethoxylines dans la Technique Moderne», Vortrag gehalten am 7. 5. 53 im Centre de Perfectionnement Technique, «Maison de la Chimie», Paris, publiziert in «Industrie des Plastiques Modernes», Juni 1954, Nr. 6.

are individual talents that can be put to use with great profit particularly as regards finding new uses for new materials and furthering technological progress. See also the article by the same author in "Allgemeine Schweizerische Militärzeitschrift", October 1959, entitled "Science and technology in the service of national defence", in which he writes: "The person who is thoroughly at home in two distinct fields of knowledge is frequently able to provide the decisive incentive."
On the occasion of the German Plastics Congress held at Stuttgart in October 1954, Dr. Rudolf Gäth, Head of the Plastics Applications Department of Badische Anilin- und Sodafabrik (BASF), delivered a paper entitled "Where do we stand today as regards the development of plastics?", in which he stated: "...a simultaneous familiarity with both metals and plastics is of inestimable value and capable of ensuring further progress".

[38] Among the many tasks of AIAG's Research Institute at Neuhausen was the testing of the huge numbers of new products constantly being brought onto the market by industry as a whole (i.e. including the chemical industry with its colorants, paints and varnishes, etc.), to determine whether they could render useful service to the aluminium industry. This work, which was first and foremost carried out as a form of customer service, also included the closer investigation of products and methods that might be detrimental or prejudicial to the aluminium industry. At that time Neuhausen had not yet tested any load-transmitting synthetic resin-based metal-bonding agents.

[39] Concerning the special importance attaching to the durability of the bond (under conditions of fatigue, etc.), see R. Gäth (Head of the Plastics Applications Department of BASF, Ludwigshafen/Rhein), "Basic principles governing the choice of a plastic"), in "Kunststoffe", Munich, October 1963; this article, which is a résumé of a paper delivered to the Plastics Congress held at Vienna on 19 April 1963, contains a section entitled "The importance of long term resistance characteristics".

[40] Compare "Chemical Processing", August 1955, pp. 6 and 7, where there is a report on an address delivered on the occasion of the presentation to Mr. Williams (Vice-President and Adviser on Research and Development to E. I. Du Pont de Nemours & Co.) of the Perkin Medal of the American Section of the Society of Chemical Industry: "Research from a management point of view. For top returns on research dollars, management must: Guard against premature selling of the lab's new products. Maintain individual initiative along with smooth relationships within groups.... Discovery and invention are highly individual matters, whereas the great mass of technical effort is concerned with the elaboration of ideas. We all know that a single concept by an individual can change the whole course of an industry. The research worker has a potential that is, without exaggeration, extraordinary."

[41] See Dr. Mario Pedrazzini, university lecturer and co-author of the standard commentary on the Swiss patent laws (Blum-Pedrazzini: "Das schweizerische Patentrecht"), lecture delivered to the Zurich Association of Jurists under the title "Rigidity and elasticity in the protection of industrial property"; reported in "Neue Zürcher Zeitung", No. 4470, 14 November 1962: "Practical considerations have led to the recognition of new forms that an invention may take, namely the so-called 'invention of function'...."

[42] Since the Netherlands was the first country in which the invention was subjected to a rigorous scrutiny, the granting of the patent assumed special importance. The proceedings before the Appeals Tribunal in The Hague took place on 17 October 1949, CIBA Limited of Basle being represented by Dr. Preiswerk.

[43] See "Sheet Metal Industries", September 1949 (London), where the report of the Swiss Federal Laboratory for Testing Materials and Research (Professor M. Ros and H. Kühne) was made available to English-reading specialists under the title "A new bonding resin, detailed results of laboratory tests on ARALDITE". At a later date Kühne, reviewing a book by E. Krekeler, published in 1956 in Cologne, "Das Verbinden von Metallen durch Kunstharzkleber" (Report No. 246 of the North Rhine-Westphalia Ministry of Economic Affairs and Transport), noted that the author's list of references failed to include any mention of the extensive work done in Switzerland on what was virtually an identical subject, including the article in "Sheet Metal Industries" referred to above (see "Schweizer Archiv für Angewandte Wissenschaft und Technik", March 1958, p. 94). Evidently the Swiss work on this subject had escaped the attention of the German-speaking world. In connection with investigations into the so-called "joint factor", which had been identified by N. A. de Bruyne (see, for example, "The strength of glued joints" in "Aircraft Engineering", Vol. XVI, 115, 1944), K. Frey used the findings obtained by the Swiss Federal Laboratory for Testing Materials and Research to compute some further laws applicable to simple bonded overlapping sheet metal specimens ("Schweizer Archiv", No. 2, Vol. 19, 1953, article entitled "The breaking strength of metal composites bonded with synthetic resins").

«The Role of Ethoxyline Resins in Modern Technology», Vortrag gehalten am 5th International Mechanical Engineering Congress, Turin, Herbst 1953. Publiziert als Druckexemplar, das an die Kongressteilnehmer verteilt wurde.
K. Meyerhans, «Äthoxylinharze in der Hochspannungstechnik», «Kunststoffe», München, Heft 10, Okt. 1953.
F. K. Trietsch, «Neuzeitliche Bindemittel und das Kleben von Metallen», «Konstruktion», Berlin, 6. 1954, Heft 4. (Nach einem Vortrage vor dem Arbeitskreis Berlin der Adki am 8. 1. 54). «Glasgewebeschichtstoffe und ihre Verwendungsmöglichkeiten», «Konstruktion», Berlin, 6. 1954, Heft 11.
E. Preiswerk, «Bindemittel und Giessharze auf Äthoxylinharzbasis (ARALDIT-Harze) in der modernen Technik», Vortrag gehalten am 29. 1. 54 vor dem Schweiz. Verbande für die Materialprüfung der Technik (SVMT) und der Schweiz. Vereinigung der Lack- und Farbentechniker (SVLFC) an der Eidg. Technischen Hochschule in Zürich. Der Vortrag fand gemeinsam mit Dr. Narracott der SHELL Petroleum Co. Ltd., London, statt, welcher über das Thema «The Place of Epoxide Resins in the Surface Coatings Industry» sprach. Über die Tagung referierte eingehend Prof. M. Hochweber, EMPA Zürich, in der «Neuen Zürcher Zeitung», 3. 2. 54, Blatt 5, Nr. 264.
K. Meyerhans, «Werkzeuge aus Kunstharzen», «Kunststoffe», Okt. 1955.
«Some New Applications of Epoxy Resins on the Continent», «The Plastics Institute Transactions», London, Okt. 1957 (als Beitrag zu einem Symposium über «Recent Continental Developments in Plastics Materials», 27. 3. 57 in London).
Rolf Müller, «Epoxyharze – Verarbeitung und Anwendung in der Industrie», «Technische Rundschau», Bern, Nr. 17, 19. 4. 63.
R. Stierli, «Epoxydharze für die Elektroindustrie», «Kunststoffe», München, Aug. 1963 (nach einem Vortrag auf dem Kunststoffkongress Wien, April 1963).

[46] Vergleiche S. 3 «ARALDIT» des anlässlich der Mustermesse 1946 an die Interessenten verteilten Faltprospektes «CIBA Kunststoffe».

[47] Die sog. «Formulators» haben im letzten Jahrzehnt – besonders in der USA – eine wachsende Bedeutung erlangt. In der Untersuchung «Epoxy Resins, Market Survey and User's Reference», HARVARD University, 1959, wird im Kapitel «Formulators» darüber ausgeführt (S. 31):
«... There are over 100 companies in the formulation of epoxy resin in the United States today and they range in size from companies as large as the Minnesota Mining and Manufacturing Company to companies with only a few employees... The production operation performed by the formulator or compounder, as he is sometimes called, starts with the purchase of raw epoxy resin from one of the several manufacturers. He also buys or sometimes manufactures curing agents and various diluents and he mixes to order a given epoxy formulation. The particular formulation he mixes on any occasion may be a more or less standard variety that was developed and recommended by one of the resin manufacturers or it may be a formulation that he developed himself. The equipment needed to make these formulations is minimal, consisting of standard mixing and weighing equipment»... Über systematische Untersuchungen in jüngster Zeit auf diesem Gebiete vergleiche u. a.: F. Hirsch und F. Koved (Aerospace Group, General Precision Inc., Pleasantville, N. Y.) im Aufsatze «Thermal conductivity of epoxy resin systems» («Modern Plastics», Okt. 64) oder R. Schmid (CIBA AG, Basel) im Vortrage «Einfluss von Füllstoffen und Additiven auf die Eigenschaften gehärteter Epoxydharze» («Neue Zürcher Zeitung», Nr. 4811 (11. 11. 64) Blatt 6).

[48] Vergleiche W. Fisch in «Chemie und Technologie der Kunststoffe» (Houwink-Staverman), 4. Auflage, Bd. II, S. 1016–1019 und S. 1025–1033.

[49] Vergleiche dazu: Untersuchung «Epoxy Resins» der Harvard University, 1959, (loc. cit., Anm. 47), S. 166/167 («Development of New Materials»).
W. Fisch in «Chemie und Technologie der Kunststoffe», 1963, (loc. cit., Anm. 48), S. 1022/1024.
R. Stierli in «Epoxydharze für die Elektroindustrie», 1963, (loc. cit., Anm. 45). Vergleiche auch den Bericht darüber in «Modern Plastics», Sept. 1963, S. 174. Siehe auch Anm. 65.

[50] Vergleiche dazu: O. Ernst und U. Niklaus, CIBA Basel, im Aufsatz «Neuentwicklungen auf dem Epoxydharzgebiet», VDI-Bericht Nr. 65, 1962, S. 79 mit Bild 4: «Zugscherfestigkeit des Versuchsproduktes X200/8766 mit Härter 903 bei höheren Prüftemperaturen». Siehe auch H. Batzer in CHIMIA, 16 (1962), S. 57.

[51] Vergleiche E. Reimer, «Patentgesetz und Gebrauchsmustergesetz», München, 1958, S. 192 (Pat. Gesetz § 6, Anm. 4: «Die Aufgabe des Patentamtes und des Gerichtes: Berücksichtigung von Gerechtigkeit und Rechtssicherheit.»). Ausführungen über die Gründe, welche die ideale Fassung einer Patentschrift oft nicht erreichen lassen. Vgl. dazu z. B. auch den solche Fragen treffend kennzeichnenden redaktionellen Aufsatz «What's in a name?» in «Mod. Plast.». Aug. 1964, S. 85.

[44] In the autumn of 1948 Dr. Preiswerk visited the United States, where he had an opportunity of discussing the findings already made in Europe with representatives of technology and industry (in particular electrical firms, Air Material Command, Dayton, Ohio, and other bodies). Soon afterwards, in 1949, the first articles on the subject were published in the U.S.: (a) "Ethoxylines, a new group of triple function resins", "Electrical Manufacturing", July 1949 (Preiswerk and Meyerhans), and (b) "New resins provide practical bonding agent for metals", "Materials and Methods", October 1949 (Preiswerk, Meyerhans, and Denz).

[45] The following is a short list of some of the articles recording the subsequent development of the resins:
E. Preiswerk, "Ethoxyline resins, observations on development up to the present – Aspects for future applications", paper delivered in September 1951 to the Summer School of Aero-Research Ltd., now CIBA (ARL) Ltd., in Cambridge, England; this paper was reproduced in the book "Structural Adhesives", publ. by Lange Ltd. in London, December 1951; "Light Metals", London, November 1951; and "Plastics", London, January/February 1952. This paper was also delivered in French on the occasion of the Jornadas Nacionales de Plásticos in Madrid in May 1952, and was published in "Revista de Plásticos", No. 16, July/August 1952, under the title "Nuevos Desarrollos en el campo de las Resinas Etoxilínicas".
K. Meyerhans, "ARALDITE-based bonding agents and casting resins", "Kunststoffe", Munich, November 1951. "Application and use of ARALDITE as bonding agent and as casting resin", "Kunststoffe", Munich, December 1951 (account of a paper delivered to the Plastics Congress held at Wiesbaden in October 1951).
E. Preiswerk, "Ethoxyline resins in electrical engineering", paper delivered on 17 October 1952 to the VDE "Plastics in Electrical Engineering" Conference held during the German Plastics Fair at Düsseldorf in 1952, published in "ETZ", Vol. 5, Ed. B, No. 1 (21 January 1953). Also published in expanded form as supplement. (A similar paper covering much the same ground was read on 16 October 1952 to the VDI, Frankfurt am Main section, under the title "The use of ethoxyline resins in modern technology".)
"Le rôle des résines ethoxylines dans la technique moderne", paper delivered on 7 May 1953 at the Centre de Perfectionnement Technique, "Maison de la Chimie", Paris, published in "Industrie des Plastiques Modernes", June 1954, No. 6.
"The role of ethoxyline resins in modern technology", paper delivered to 5th International Mechanical Engineering Congress, Turin, autumn 1953. Printed in the form of a pamphlet and distributed to those attending the congress.
K. Meyerhans, "Ethoxyline resins in high tension engineering", "Kunststoffe", Munich, No. 10, October 1953.
F. K. Trietsch, "Modern bonding agents and the bonding of metals", "Konstruktion", Berlin, Vol. 4 (account of a paper delivered to the Berlin section of ADKI on 8 January 1954).
"Glass fabric laminates and their application potential", "Konstruktion", Berlin, Vol. 6, 1954, No. 11.
E. Preiswerk, "Ethoxyline (ARALDITE) resin-based bonding agents and casting resins in modern technology", paper delivered on 29 January 1954 to the Swiss Association for Technological Materials Testing and the Swiss Paint and Varnish Manufacturers Association at the Federal Institute of Technology, Zurich. At the same meeting a paper entitled "The place of epoxide resins in the surface coatings industry" was read by Dr. Narracott of the Shell Petroleum Co. Ltd., London. A detailed report of the meeting was given in the "Neue Zürcher Zeitung", No. 264, p. 5, 3 February 1954, by Professor M. Hochweber, of the Swiss Federal Laboratory for Testing Materials and Research.
K. Meyerhans, "Plastic tools", "Kunststoffe", October 1955.
"Some new applications of epoxy resins on the Continent", in "The Plastics Institute Transactions", October 1957 (contribution to a symposium on "Recent Continental Developments in Plastics Materials", London, 27 March 1957).
Rolf Müller, "Use and applications of epoxide resins in industry", "Technische Rundschau", Berne, No. 17, 19 April 1963.
R. Stierli, "Epoxide resins in the electrical industry", "Kunststoffe", Munich, August 1963 (account of a paper read to the Plastics Congress held at Vienna in April 1963).

[46] Compare page 3, "ARALDITE", of the brochure "CIBA Plastics" that was distributed to visitors to the CIBA stand at the 1946 Swiss Industries Fair.

[47] Formulators have acquired increasing importance over the last ten years, particularly in the United States. In "Epoxy Resins, Market Survey and User's Reference", Harvard University Press, 1959, the following remarks are made in the section on "Formulators" (p. 31): "... There are over 100 companies in the formulation of epoxy resin in the United States today and they range in size from companies as large as the Minnesota Mining and Manufacturing Company to companies with only a few employees.... The production operation performed by the formulator, or compounder as he is sometimes called, starts with the purchase of raw epoxy resin from one of the several manufacturers. He also

[52] Vergleiche z. B. den Inseratentext der Rubber and Asbestos Corp. in «Modern Plastics», März 1959, S. 195: «...This «100%-solids» adhesive flows like a heavy-bodied motor oil... Don't let the «100%-solids» term confuse you... this modified epoxy adhesive is a free-flowing liquid!»... Auch wenn das Harz in Form von Pulvern, Folien usw. zum applikatorischen Einsatz kommt, so durchläuft es während des Härtungsprozesses – auf jeden Fall zum Teil – einmal die Phase der «free-flowing liquid».

[53] Vergleiche dazu: «Modern Plastics Encyclopedia for 1960», S. 730 ff.: Aufsatz «Cast and Casting Thermosets» by M. Hilrich: «...casting resins, while bracketed as liquids, actually range from... liquids... to puttylike compounds... as laminating, splining, trowelling, to make tools... as adhesives to create durable bonds... most of the special property casting resins are in the epoxy family...» «epoxy adhesives have made great strides because they solve joining problems where conventional adhesives are unsatisfactory»...
Siehe auch S. 99 der «Epoxy Resins, Market Survey» (loc. cit., Anm. 47): «...Those instances in which epoxies are used strictly as an adhesive are difficult to isolate»... «Epoxies are used... in place of solder»...

[54] Es ist selbstverständlich, dass sich der neue «Metallbinder» bei seinem Erscheinen in der Technik zuerst einmal in die Klasse der herkömmlichen (conventional) Klebmittel (adhesives) einreihen musste, obwohl er gerade den Rahmen derselben sprengte. – «A revolution is taking place in the arts of joining and fastening. The resinous adhesives are replacing welding, soldering, brazing, rivets and even nails, and the epoxies are leading the way», schreibt I. Skeist einleitend zu seinem Kapitel 10 («Adhesives, Caulking and Patching») seines 1958 erschienenen Handbuches «Epoxy Resins» (New York).

[55] In der «Internat. Zeitschrift für das gesamte Kunststoffgebiet» «Kunststoffe-Plastics», 5, Nr. 1, 1958, schrieb Fritz Ohl im Aufsatz «Kunststoffe als Klebmittel» zu dieser Frage: «...es ist auch nahezu unmöglich, die Kunststoffe in ihrer Verwendung als Klebmittel gruppenmässig oder anwendungstechnisch exakt zu klassifizieren. Bereits mit den beiden DIN-Blättern 16920/16921 wurde erfolgreich angestrebt, Klarheit in das ebenso umfangreiche wie komplizierte Gebiet der Klebmittel zu bringen»...
In Ullmanns «Encyclopädie der Technischen Chemie», Bd. 9, S. 564 ff., beleuchteten G. Schulz und H. Beuschel im Kapitel «Kitte, Verguss und Dichtungsmassen» die Situation mit folgenden Worten: «...Kitte, Verguss und Dichtungsmassen sind zur Füllung von Löchern, Rissen, Sprüngen oder Vertiefungen und zur dauerhaften Vereinigung einzelner Teile bestimmt. Die drei Begriffe sind nicht klar voneinander zu trennen und gehen vielfach ineinander über. Zwischen Kitten und Klebstoffen besteht eine gewisse Verwandschaft, da sie Haftvermögen auf anderen Werkstoffen besitzen. Auch hier ist eine scharfe Trennung oft schwierig, so dass gewisse Überschneidungen mit dem Stichwort Klebstoff nicht zu vermeiden sind»... Bei der anschliessenden Behandlung der chemischen Stoffklassen (welche als Kitte, Verguss- und Dichtungsmassen in Betracht kommen), wird unter dem Abschnitt «Epoxyharze» als einziges und wesentliches vermerkt (S. 570): «Epoxyharze werden benutzt zur Verbindung von Metallen, insbesondere Leichtmetallen, aber auch Buntmetallen... oder mit keramischen Massen, Holz, Glas oder Faserstoffen... Die Epoxyharze sind in Form von Flüssigkeiten, Pasten, Stangen oder Pulver im Handel...»

[56] In den USA definierte R.F. Blomquist, Forest Prod. Lab., U.S. Dep. of Agriculture, in der «Encyclopedia of Chem. Technology», First-Suppl. Vol., New York 1957, im Aufsatze «Adhesives», S. 18, die Verhältnisse wie folgt: «...Current usage defines an adhesive as a substance capable holding materials together by surface attachment. The term «adhesives» is now considered as a general term that includes other materials such as cements, glues etc.... Although all of these terms are loosely used interchangeably, the term «Adhesive» is generally becoming most widely used and it is considered the most acceptable general term for all such bonding agents. Adhesives are commonly subdivided on the basis of several factors such as their application temperature... their end use (laminating adhesives, assembly adhesives)...»

[57] In seiner «Technologie der Klebstoffe» (loc. cit., Anm. 4) schrieb C. Lüttgen, Mitglied des Deutschen Patentamtes im Teil 1 im Jahre 1959 einleitend: «...Unter dem Begriff «Klebstoffe» fasst man alle diejenigen Stoffe zusammen, die geeignet sind, Gegenstände aus gleichem oder verschiedenem Material miteinander zu verbinden. Im Sprachgebrauch gibt es viele Ausdrücke für Klebstoffe, wie Leim, Kleister, Binder, Zement, Kitt usw., ohne dass man unter diesem Namen genau definierte Produkte versteht... Heute ist es so, dass fast jeder Berufszweig eine andere Bezeichnung für die von ihm verwendeten Klebstoffe geprägt hat und dass diese Bezeichnungen nicht einmal immer das gleiche bedeuten. Der Begriff «Klebstoff» umfasst jedoch das gesamte Gebiet der zum Leimen, Kleben, Kitten und Binden verwendeten Stoffe.»... Im Teil 2 (1. Auflage, 1957), hielt der Autor u.a. fest: S. 70: «...Bei den mit Hilfe von härtbaren Harzen hergestellten Kitten handelt es sich

buys or sometimes manufactures curing agents and various diluents and he mixes to order a given epoxy formulation. The particular formulation he mixes on any occasion may be a more or less standard variety that was developed and recommended by one of the resin manufacturers or it may be a formulation that he developed himself. The equipment needed to make these formulations is minimal, consisting of standard mixing and weighing equipment...." For information concerning systematic investigation carried out in recent years in this field see, for example, F. Hirsch and F. Koved (Aerospace Group, General Precision Inc., Pleasantville, N.Y.) in the article "Thermal conductivity of epoxy resin systems", "Modern Plastics", October 1964, or R. Schmid (CIBA Limited, Basle), "Influence of fillers and additives on properties of cured epoxide resins", "Neue Zürcher Zeitung", No. 4811 (11 November 1964), p. 6.

[48] See W. Fisch in "Chemie und Technologie der Kunststoffe" (Houwink-Staverman), 4th Ed., Vol. II, pp. 1016–1019 and pp. 1025 to 1033).

[49] See also:
Epoxy resins survey, Harvard University Press 1959 (loc. cit., Note [47] above), pp. 166 and 167 ("Development of New Materials").
W. Fisch in "Chemie und Technologie der Kunststoffe", 1963 (loc. cit., Note [48] above), pp. 1022–1024.
R. Stierli in "Epoxide resins in the electrical industry" 1963 (loc. cit., Note [45] above). See also the report in "Modern Plastics", September 1963, p. 174, and see Note [65] below.

[50] See O. Ernst and U. Niklaus (CIBA Limited, Basle), in their article "New developments in the field of epoxide resins", VDI Report No. 65, 1962, p. 79, with illustr. 4: "Tensile shear strength of trial product X200/8766 with hardener 903 at elevated temperatures." In addition see H. Batzer in "Chimia", Vol. 16 (1962), p. 57.

[51] See E. Reimer, "Patentgesetz und Gebrauchsmustergesetz", Munich 1958, p. 192 (Law on Patents § 6, note 4: "Functions of patent office and court: consideration of rights and legal protection"): an account of the reasons why it is frequently difficult to arrive at the ideal wording for a patent specification. See also the editorial "What's in a name?" in "Modern Plastics", August 1964, p. 85.

[52] Compare the text of an advertisement by the Rubber and Asbestos Corp. in "Modern Plastics", March 1959, page 195: "This '100%-solids' adhesive flows like a heavy-bodied motor oil.... Don't let the '100%-solids' term confuse you...this modified epoxy adhesive is a free-flowing liquid!" Even when the resin is applied in the form of powder, foil, etc., it passes, at least in part, through the "free-flowing liquid" stage at some point during the curing process.

[53] Compare "Modern Plastics Encyclopedia for 1960", pp. 730 ff., article on "Cast and casting thermosets" by M. Hilrich: "...casting resins, while bracketed as liquids, actually range from... liquids...to putty-like compounds... for laminating, splining, trowelling, to make tools...as adhesives to create durable bonds ...most of the special property casting resins are in the epoxy family.... Epoxy adhesives have made great strides because they solve joining problems where conventional adhesives are unsatisfactory...." See also p. 99 of "Epoxy Resins, Market Survey" (loc. cit., Note [47] above): "...Those instances in which epoxies are used strictly as an adhesive are difficult to isolate.... Epoxies are used...in place of solder...."

[54] For obvious reasons the new "metal-bonding agent" was classified as a conventional adhesive when it first appeared, although in fact it was anything but conventional. "A revolution is taking place in the arts of joining and fastening. The resinous adhesives are replacing welding, soldering, brazing, rivets and even nails, and the epoxies are leading the way", writes I. Skeist in his introduction to Chapter 10 ("Adhesives, caulking, and patching") of his manual "Epoxy Resins" (New York, 1958).

[55] In "Kunststoffe-Plastics", Vol. 5, No. 1, 1958, Fritz Ohl writes, in his article "Plastics as adhesives": "It is almost impossible to devise a precise method for classifying plastics used as adhesives on the basis of either their general category or their mode of application. The two DIN [German Industrial Standards] specifications 16920 and 16921 represent a successful attempt to introduce order into the field of adhesives, which is as wide in scope as it is complicated."
In Ullmann's "Encyclopädie der Technischen Chemie", Vol. 9, pp. 564 ff., G. Schulz and H. Beuschel, in the chapter entitled "Cements, casting resins, and sealing compounds" explain the situation as follows: "...Cements, casting resins, and sealing compounds are designed for filling holes, cracks, fissures, and unevennesses and for providing a durable bond between parts. It is difficult to define the borderlines between the three terms, and they overlap to a considerable extent. There is a certain relationship between cements and adhesives, since they both exhibit the property of adhering to other materials. Here again the exact borderline is difficult to delineate, and it is impossible to avoid a certain overlapping with the class of substances termed 'adhesives'." There follows a discussion of the chemical classes in-

letzten Endes um die gleichen Produkte, die bereits im ersten Bande ausführlich als Klebstoffe lediglich durch die pastenförmige Konsistenz...» und S. 75: «...Schnellhärtender Kitt mit dielektrischen Eigenschaften zum Verkitten von Transformatoren, Kondensatoren usw....»

[58] Ob in der Praxis die Räume zwischen den zu verbindenden Flächen «dünn» oder «dick» sind, so werden zur Messung der Bindefunktion mit Vorteil die einfach herzustellenden, leicht reproduzierbaren und übersichtlichen Scherfestigkeitsproben an «verklebten» Metallblechen mit einheitlich dünner Fuge durchgeführt.

[59] Vergleiche dazu z.B. loc. cit. Anm. 5 mit der Auslegeschrift der Firma Henkel, wo beim «Vergiessen von Hohlräumen, besonders in elektrischen Apparaten mittels Polyepoxydverbindungen» die «Dauerfestigkeit der Klebverbindungen von besonderer Bedeutung» ist. In entsprechender Weise formuliert Dr.-Ing. Karl Mienes in seinem Buch «Kunststoffverarbeitung» (ECON-Verlag, Düsseldorf 1955) im Kapitel II «Verarbeiten von ungeformten Kunststoffen in festem, flüssigem, gelöstem oder dispergiertem Zustand» auf S. 38: «...Mit ungesättigten Polyestern gemachte Erfahrungen haben gezeigt, dass deren Anwendung in manchen Fällen über den vielfach benutzten Sammelbegriff «Giessharz» hinausgeht. Gerade bei den Äthoxylin-(Epoxy-)Erzeugnissen[1] ist klarzustellen, dass hier die Giessharze zumeist bereits eine Bindemittelfunktion kraftschlüssigen Charakters erfüllen. Die Hersteller dieser Harze[2] formulieren den damit erzielten technischen Fortschritt in der Weise, dass ARALDIT-Harze das Verbinden und Vereinigen von Werkstoffen verschiedenster Art in einem mechanisch und elektrisch hochwertigen Verbande[3] gestatten. Hohe Bindefähigkeit und Güte, vermeidbares Schrumpfen und beim Härten sich erübrigender Pressdruck – das sind erfolgbringende Gesichtspunkte für technisch verwendete Äthoxylinharze.»
Ähnlich schreibt z.B. Dr. E. Plath (Forschungsinstitut für Holzwerkstoffe und Holzleime, Karlsruhe) im «Taschenbuch der Kitte und Kunststoffe», Wissenschaftliche Verlagsgesellschaft, Stuttgart, 1963 (S. 314/315) im Abschnitt «Epoxydharze»: «...Elektromaschinen... wobei sich die Verwendungsgebiete von Kleb- und Giessharzen etwas überschneiden»...
Bemerkenswert ist auch die Feststellung von Dipl.-Phys. A. Rost, Essen, im Aufsatz «Verarbeitung von Epoxyd-Giessharzen» (VDI-Bericht Nr. 65, 1962): ...«Eine Giessharzmasse braucht also nach dem Normentwurf 16945 nur giessbar zu sein. Ihr Anwendungsgebiet wird dadurch nicht berührt»... (Vergleiche dazu Anm. 61).

[60] Vergleiche dazu H. Stäger in Houwinks «Chemie und Technologie der Kunststoffe», Band II, 2. Auflage, Leipzig 1942, S. 31 und 33: «Die Giessharze. Die Giessharze sind auch Edelkunstharze genannt, weil sie als Ersatz für bestimmte edle Schmuckgegenstände benutzt werden. Es war schon seit Jahren das Bestreben verschiedener Kunstharzfachleute, Phenolformaldehydharze zu vergiessen, um auf diesem Wege direkte Formstücke zu erhalten...... neuerdings die Giess- oder Edelkunstharze bereits eine grosse Bedeutung erlangt... so sollen sie in den Vereinigten Staaten von Amerika auf dem Gebiete der Schnallen und Knöpfe langsam Kasein, Horn, Bein und Perlmutter aus dem Felde schlagen... jetzt schon Besteckgriffe, Utensilien für Lampen, Rauchtische, Möbel und Autobestandteile in grossen Mengen hergestellt. Dieser Verbrauch dürfte ungefähr 75% des Gesamtumsatzes an Giessharzen ausmachen. Neben der Hauptanwendung der Edelkunstharze an Stelle von Horn, Bein, Elfenbein, Bernstein, Kunsthorn usw. sind auch noch einige technische Verwendungszwecke zu erwähnen: Es werden gewisse säurebeständige Bestandteile für die Kunstseidenerzeugung daraus hergestellt, für die Elektrotechnik werden Stützisolatoren für 10–150 kV Betriebsspannung bei Innenaufstellung erzeugt»...
Vergleicht man diesen durch Stäger geschilderten, damaligen Stand der Technik betr. Verwendungsmöglichkeiten für Giessharze mit den Angaben, welche in dem von de Trey (Dr. Castan) im Jahre 1940 veröffentlichten Schweizer Patent Nr. 211 116 für die Anwendung der Äthoxylinharze zu finden sind, so erkennt man unschwer, welcher Art damals die allgemeine Blickrichtung für die Applikation eines sog. «Giessharzes» war!

[61] Vergleiche Deutsche Normen, DIN 16945 (Entwurf Jan. 1962), «Prüfung von Giessharzen», Begriffe: «Giessharze sind flüssige oder schmelzbare... Harze, die für sich oder mit Reaktionspartnern (Härtern) ...in einen vernetzten Zustand übergehen.»

[62] Vergleiche z.B. «Encyclopädie der elektrischen Isolierstoffe» (CES/CEI), 1. Auflage, Zürich 1960. Unter dem Stichwort «Giessharze» werden in der Gruppe 522 die «Giessharze» den «Imprägnierharzen» gleichgestellt.

---

[1] Über Äthoxylinharze s. a. K. Frey, Chimia, 8, 1954, S. 1–6, E. Preiswerk, Plastics, Jan./Febr. 1952, ferner J. G. Carey, Modern Plastics 30, 1953, S. 130–134.
[2] Ciba Aktiengesellschaft, Basel, Schweiz.
[3] Über ARALDIT als Konstruktionselemente s. a. E. Preiswerk, Plastverarbeiter 2, 1954, S. 43–45.

---

volved, and in the section on "Epoxy resins" the only information of substantial importance is that given on p. 570: "Epoxy resins are used for the bonding of metals, particularly light alloys but also non-ferrous metals...or ceramics, wood, glass, and fibres.... Epoxy resins are marketed in the form of liquids, pastes, rods, or powders...."

[56] R. F. Blomquist (Forest Products Lab., U.S. Dept. of Agriculture), has described the situation as follows in the "Encyclopedia of Chemical Technology", 1st Suppl. Vol., New York 1957, in the section on "Adhesives" (p. 18): "...Current usage defines an adhesive as a substance capable of holding materials together by surface attachment. The term 'adhesives' is now considered as a general term that includes other materials such as cements, glues, etc. Although all of these terms are loosely used interchangeably, the term 'adhesive' is generally becoming most widely used and it is considered the most acceptable general term for all such bonding agents. Adhesives are commonly subdivided on the basis of several factors such as their application temperature...their end use (laminating adhesives, assembly adhesives)...."

[57] In Part 1 of his "Technologie der Klebstoffe" (loc. cit., Note [4] above), C. Lüttgen introduced his subject (1959) as follows: "The term 'adhesives' embraces all substances that are used for joining objects made of similar or dissimilar materials. In common usage there are numerous terms for different types of adhesive substance, such as glue, gum, binder, paste, cement, bonding agent, and many others, but they do not accurately define any single substance or group of substances: in fact virtually every trade or profession uses its own peculiar terms, and these terms do not always agree. Nevertheless, the term 'adhesives' can be regarded as covering all substances that are used for gluing, cementing, and bonding." In part 2 of his book (1st Ed., 1957), Lüttgen says (p. 70): "The cements that are made with the aid of curable resins are in fact the same as those that have already been dealt with in volume one as adhesives...." On page 75 Lüttgen continues: "Rapid-drying compound with dielectric properties for embedding transformers, capacitors, etc."

[58] Whether in actual practice the gap between the surfaces to be united is large or small, measurement of the bonding function is most conveniently carried out by using overlapping bonded metal strips with a narrow gap of uniform size, since shear strength tests carried out in this manner are reproducible and reliable.

[59] See the Henkel patent specification (loc. cit., Note [5] above), where it is stated that "the durability of the bond" is of special importance for "the filling of hollow spaces, particularly in electrical apparatus...with polyepoxide compounds". Similarly Dr. Karl Mienes in his book "Kunststoffverarbeitung" (ECON-Verlag, Düsseldorf 1955), Chap. II, "Use of plastics materials in solid, liquid, dissolved, or dispersed state", on page 38: "Experience with unsaturated polyesters has shown that in many cases their use goes beyond the normal scope understood by the commonly used term "casting resin". In the particular case of the ethoxyline (epoxy) resins[1], we are dealing with a casting resin that simultaneously performs a bonding function. The producers of these resins[2] sum up the situation by stating that the technical progress they represent is due to the fact that ARALDITE may be used for the uniting and bonding of a wide range of dissimilar materials to form a composite structure of first-class mechanical and electrical properties[3]. High bond strength and end product quality, negligible shrinkage and the ability to cure without the need of pressure, are the factors that have assured the prosperity of the ethoxyline resins." Similar opinions are expressed by Dr. E. Plath (Karlsruhe Research Institute for Woods and Wood Glues) in his "Taschenbuch der Kitte und Kunststoffe" (Wissenschaftliche Verlagsgesellschaft, Stuttgart 1963), section "Epoxydharze" (pp. 314–315): "Electric machines in which the functions of adhesive resins and casting resins overlap".
A. Rost (Essen), in his article "Use of epoxy casting resins" (VDI Report No. 65, 1962), also has an interesting point to make in this respect: "According to the draft Standard 16945 a casting resin merely has to be pourable; this does not affect its field of application." See also Note [61] below.

[60] See H. Stäger in Houwink's "Chemie und Technologie der Kunststoffe", Vol. II, 2nd Ed., Leipzig 1942, pages 31 and 33: "Casting resins. In the German language casting resins are also known as 'Edelkunstharz', literally 'precious synthetic resins', since they are used as a replacement for various other ornamental materials. For many years plastics specialists have been developing pourable phenolformaldehyde resins so that shaped parts could be cast directly. Recently these casting resins have assumed considerable importance: thus in the United States they are gradually ousting casein horn, bone, and mother-of-pearl for the production of buckles and buttons, are also being used for cutlery handles, lamp parts, smokers' requisites, furniture, and automobile accessories.

---

[1] cf. K. Frey, "Chimia", 8, 1954, pp. 1–6, E. Preiswerk, "Plastics", Jan./Feb. 1952, and J. G. Carey, "Modern Plastics", 30, 1953, pp. 130–134.
[2] CIBA Limited, Basle, Switzerland.
[3] E. Preiswerk, "Plastverarbeiter", 2, 1954, pp. 43–45.

[63] Vergleiche Deutsche Norm, DIN 7708 (Okt. 1957), Begriff «Formstoff.» «Formstoffe sind Stoffe, die aus Formmassen... durch spanlose Formung... hergestellt worden sind und die dann als Formteil oder Halbzeug vorliegen.» – Entsprechend siehe Deutsche Norm 16945 (loc. cit. Anm. 61) «Giessharz-Formstoffe sind Werkstoffe aus gehärteten Giessharzmassen.» Die Herstellung von reinen (oder nur «gefüllten») Äthoxylinharz-Formstücken oder Halbzeug, bei welchen das Kleben, die Bindefunktion praktisch von keiner Bedeutung sind, tritt auch in der heutigen Technik – im Vergleich zu anderen Kunststoffen – in relativ nur geringem Maßstabe auf (vgl. hiezu auch Anm. 4 und 84). In dieser Beziehung haben die Äthoxylin-(Epoxy-)Harze gegenüber den bestehenden, preisgünstigeren Phenol-, Stryrol-, Anilin- oder Polyesterharzen grundsätzlich nichts Neues gebracht, wenn sich auch mit innen Gußstücke von erheblicher Grösse des Einzelstückes, guter Masshaltigkeit und geringer Rissanfälligkeit herstellen lassen. (Über die diesbezüglichen Entwicklungstendenzen auf dem Gebiete der Polyesterharze siehe z.B. Rembold und Keller, MICAFIL AG, Zürich, «Neuartige Katalysatoren eröffnen Polyester-Giessharzen interessante Möglichkeiten» in «Kunststoffe», Sept. 1964, S. 554ff).

[64] Im März 1954 konnte F. Vinsonneau, technischer Direktor der Firma SNCASO schreiben (vgl. Zeitschrift «Interavia», März 1954, S. 155): «...Das Metallkleben kam schon beim ersten Prototyp des «Vautour» (001) zur Anwendung und hat sich gut bewährt. Wie wir glauben, handelt es sich um die erstmalige Erprobung dieses Verfahrens an lebenswichtigen Teilen. ...In über 100 Versuchsflügen, darunter zahlreichen Durchbrechungen der Schallmauer, wurde die Qualität des «Vautour» unter Beweis gestellt. ...Und am 29. 12. 53 hatten wir die Genugtuung, sogar unseren Luftfahrtminister Louis Christaens zum ersten Überschallflug eines aktiven Staatsmannes an Bord der «Vautour» zu beglückwünschen»... – In der U.S.-Zeitschrift «Aviation Age for Aviation's Technical Management», Dezember 1954, berichtete anschliessend der europäische Korrespondent W.P. Moser, S. 38ff. mit den Worten: «French Use New Bonding Method on Triple Threat «Vautour»»... SNCASO's enthusiasm over a production technique that contributed materially to bringing about this happy state of affairs is only natural... And there can be no doubt that plastic bonding did help the «Vautour»; in every application so far, the new production technique has led to saving in dollars. A special claim made for the plastic bonding job on the «Vautour» is that it represents the first attempt to use new method on a «vital» aircraft part... The specific plastic SNCASO settled on was ARALDITE I, an ethoxyline resin produced by the Plastics Division of CIBA Ltd., of Basle, Switzerland. ARALDITE comes in the form of sticks, as powders...». Bereits im Mai 1952 waren die «ARALDITE Adhesives» für den englischen Flugkörperbau durch Erteilung der AIRCRAFT MATERIAL SPECIFICATION D.T.D. 861 «Adhesive for Metal» («Low Pressure Type») des Ministry of Supply homologisiert worden. – Vergleiche dazu auch in «Milestones in The History of CIBA (ARL) Ltd, 1934–1959», Duxford-Cambridge 1959, S. 26: «The first aeroplane incorporating metal honeycomb in this country was the AVRO «VULCAN» and the adhesive used was ARALDITE»...
Auf die nicht minder interessanten und sehr weitgehenden Einsätze der Äthoxylin-(Epoxy-)Harze als Metallbindemittel – teilweise kombiniert mit Glasfasern – beim Bau des CONVAIR B 58 «Hustler», des CONVAIR 880/990 «Coronado», des Jindivik-Zielflugzeuges der Government Aircraft Factories (Department of Defence Production) in Melbourne, Australia, bei NAPIER's «Spraymat»-System – um nur einige zu nennen – sei hingewiesen. Beachtenswert ist z.B. auch die Kombination Glasfaser, Metall und Äthoxylin-(Epoxy-)Harz zur Herstellung höchstbeanspruchter Flugzeugpropellerblätter durch die CURTISS-WRIGHT Corp., Caldwell, New Jersey oder die reichliche Verwendung der Kombination Glasfaser, Äthoxylin-(Epoxy-)Harz und Glasfaser/Phenolharz-Honigwabenmaterial im Verkehrsflugzeug Typ 727 von BOEING. («Mod. Plast», Aug. 1964, P. 100ff.).

[65] Vergleiche W.P. Moser in «Aviation Age for Aviation's Technical Management», New York, April 1957, S. 108ff.: «Shell construction, ARALDITE Bonds give Strength to Swiss A-A Missile.» (Nach einem Aufsatz von F.B. Stencel, in «Aluminium Suisse», September 1956, S. 119ff: «Neue Leichtbauweisen für die Zelle der leitstrahlgesteuerten OERLIKON-Fliegerabwehrrakete.») Für weitere, anschliessende Entwicklungen vergleiche z.B. den Aufsatz: «Die Technik des Leichtbaues bei Flieger- und Panzerabwehrflugkörpern», H.P. Schneiter, CONTRAVES AG, Zürich, in «Flugwehrund Technik», Jan./Febr. 1964. Es wird hier über den «Klebstoff CY-175 (Ciba Basel)» berichtet, der zur Herstellung von Druckgefässen aus Stahlbändern und Stahldraht dient. Der «Klebstoff CY 175» ist das ARALDIT-Produkt CY 175 (ein cycloaliphatisches Epoxydharz, vgl. Anm. 49), über welches R. Stierli (loc. cit., Anm. 45) in «Kunststoffe», Aug. 1963 berichtet. Auf den S. 544/545 wird dort festgehalten, dass der neue Typ auch dann noch eine erhöhte – wenn auch reduzierte – Warmfestigkeit aufweist, wenn mittels reaktiver Flexibilisatoren die mechanischen Gütewerte so verbessert wurden, dass sie z.B. wieder denjenigen des klassischen

These applications are reported to account for 75% of the total prodution of casting resins. Besides the principal use of casting resins, namely as a replacement for horn, bone, ivory, amber, artificial horn, etc., certain industrial applications may also be mentioned: thus they are used for certain acid-resistant components in the rayon industry, and in electrical engineering they are used for the manufacture of pin-type insulators for working voltages of 10 to 150 kV...."
If we compare Stäger's description of the situation in 1942, and in particular the applications of "casting resins", with the data given by Castan in the de Trey patent (No. 211,116) of 1940 concerning the application of his ethoxyline resins, it is not difficult to understand what was meant in those days by the term "casting resin".

[61] Compare German Industrial Standard DIN 16945 (draft of January 1962), "Testing of casting resins", definition of terms: "Casting resins are liquid or liquefiable resins that, with or without the addition of hardeners, become cross-linked."

[62] See also "Encyclopädie der elektrischen Isolierstoffe" (CES/CEI), 1st Ed., Zurich 1960. In the entry "Giessharze", in group 522, "casting resins" are equated with "impregnating resins".

[63] Compare German Industrial Standard DIN 7708 (October 1957), definition of "moulded material": "Moulded materials are made from moulding compounds by moulding, and may be either finished moulded articles or semifinished...." See also DIN 16945 (loc. cit., Note [61] above): "Moulded materials made with casting resins are cast materials resulting from the curing of casting compounds." The production of moulded articles consisting of pure (or even "filled") epoxy resin, as compared with other synthetic resins, is today practised on only a limited scale (see also Notes [4] and [84]) in applications where adhesiveness, i.e. their bonding function, is not of immediate practical importance. In this respect epoxy resins have no special advantages over the other existing – and cheaper – phenolic, styrene, aniline, or polyester resins, even though they can be used for the production of very large castings possessing good dimensional stability and resistance to cracking. (Regarding similar developments in the field of polyester resins see also Rembold and Keller, of Micafil AG, Zurich, "New catalysts suggest interesting possibilities for polyester casting resins" in "Kunststoffe", September 1964, pp. 554 ff.)

[64] In March 1954 F. Vinsonneau, Director of Engineering of the French company SNCASO (today Sud-Aviation) wrote (see "Interavia", March 1954, p. 155): "...Here we would merely emphasise the wide use of metal bonding, which, introduced from prototype 001 onwards, has since proved its worth. We believe that this was the first time the metal bonding process was used on major structural components.... In more than a hundred test flights and several dozen passages through the 'barrier', the 'Vautour' has shown what it is worth.... And we were happy and proud on December 29th, when M. Louis Christaens, Secretary of State for Air, became the world's first Supersonic Minister aboard the 'Vautour'." In the American "Aviation Age for Aviation's Technical Management" of December 1954 W. P. Moser, the journal's European correspondent, wrote (pp. 38 ff.): "French Use New Bonding Method on Triple Threat 'Vautour'.... SNCASO's enthusiasm over a production technique that contributed materially to bringing about this happy state of affairs is only natural.... And there can be no doubt that plastic bonding did help the 'Vautour': in every application so far, the new production technique has led to a saving in dollars. A special claim made for the plastics bonding job on the 'Vautour' is that it represents the first attempt to use the new method on a 'vital' aircraft part.... The specific plastic SNCASO settled on was ARALDITE I, an ethoxyline resin produced by the Plastics Division of CIBA Ltd., of Basle, Switzerland. ARALDITE comes in the form of sticks and powders...." Already by May 1952 "ARALDITE Adhesives" had been officially standardised in the British Ministry of Supply's Aircraft Material Specification D.T.D. 861 "Adhesive for Metal (Low Pressure Type)". See also "Milestones in the History of CIBA (ARL) Ltd., 1934–1959" (Duxford, Cambs., 1959), p. 26: "The first aeroplane incorporating metal honeycomb in this country was the Avro 'Vulcan' and the adhesive used was ARALDITE."
Mention should also be made of the no less interesting and very extensive use of epoxy resins as a metal-bonding agent (in some cases combined with glass fibre) in the Convair B58 "Hustler", in the Convair 880/990 "Coronado", in the "Jindivik" target aircraft designed by the Government Aircraft Factories (Department of Defence Production) in Melbourne, Australia, and in Napier's "Spraymat" system – to name only a few instances. Special interest also attaches to the combination of glass fibre, metal, and epoxy resin for the production of heavy-duty aircraft propeller blades by the Curtiss-Wright Corp., Caldwell, N.J., and to the extensive use of the combination glass fibre, epoxy resin, and glass fibre-phenolic resin honeycomb material in the Boeing 727 ("Modern Plastics", August 1964, pp. 100 ff.).

[65] See W. P. Moser in "Aviation Age for Aviation's Technical Management", New York, April 1957, pp. 108 ff., "Shell construction, ARALDITE bonds give strength to Swiss A.-A. missile",

Äthoxylin-(Epoxy-)Harzes, dem ARALDIT-Giessharz B (dem alten Typ von de Trey) entsprechen.

[66] Im Raketenbau werden vornehmlich im Wickelverfahren («filament winding») – besonders auch unter Verwendung von Glasfasern – Schubdüsen, Brennkammergehäuse für Feststoffmotoren, Treibstoff- und Druckbehälter usw. gefertigt. Äthoxylin-(Epoxy-) Harze werden dort («Rocket cases of POLARIS, MINUTEMAN, ATLAS/ROCKETDYNE Triebwerke, TITAN rocket engine u. a. m.) eingesetzt, wo höchste Festigkeiten gefordert werden. – Ein redaktioneller Artikel «Aerospace Age» in «Modern Plastics», New York, Februar 1962, hielt dazu fest: «...The filament wound structure offers a strength-to-weight ratio that is three times more efficient than titanium... The POLARIS missile, for example, with first and second-stage motors encased in steel had a range of 1380 nautical miles; with a filament-wound epoxy-glass second stage, the range was increased to 1725 miles; and with 2875 miles as the goal, work is progressing on winding the cases of both first and second stages»...

[67] Vergleiche «Modern Plastics», Mai 1962, S. 110 ff.: Bericht über die 19th Annual Conf. der Western Sect. SPI: «...D. M. Hatch, Manager of AEROSPACE Plastics Engineering for the H. I. Thompson Fiber Glass Co. Gardena, Calif., illustrated with slides and test firing movies his spectacular account of the NOVA-scale and larger plastic nozzles for the mammoth solid propellant rockets now under investigation and funding by USAF and NASA. Hatch electrified the packed meeting room on Macrostructures with details of a proposed plastic nozzle... He also stated that filament-wound fibrous glass and epoxy solid motor cases up to 50ft. (15 Meter!) in diameter are under design and development by the plastics industry in California»...

[68] Vergleiche z. B.: «Modern Plastics», New York, Oktober 1961, Aufsatz: W. B. Royall and R. W. Matlock: «Epoxy passes toughest test: Outer space»...: «...The success of the Courier 1-B satellite will be of more than passing interest for industrial users concerned with the performance of reinforced epoxies under conditions of extreme heat and cold, intense radiation, and high vacuum. This space vehicle, still in orbit more than a year after launch, relies heavily on glass-reinforced epoxy for its structure. Its continued presence in the sky makes the material a promising candidate for more earthbound applications where service environments are severe and critical»...
«Flight International», London, 19. April 1962, S. 612: «US/UK-Satellite. Structure: Basic structural material in the satellite is plastic bonded glass-fibre. The central body section is an epoxy-bonded monofilament-wound cylinder structure; the upper dome is bonded to an aluminium ring which mates with an aluminium ring bonded to the mid skin,... aluminium cylinder... bonded to this are eight glass-fibre stiffening ribs»...

[69] Über die Bedeutung des Erfolges und der Bewährung eines neuen Werkstoffes und Verfahrens in der Flugkörpertechnik schrieb Prof. Dr. Ludwig Erhard, deutscher Bundesminister für Wirtschaft (heute Bundeskanzler) in einem Geleitwort zur 1. Nummer der Zeitschrift «Flugkörper», Febr. 1959: «Die Luftfahrtindustrie ist bekanntlich der Schrittmacher des technischen Fortschrittes, der wiederum für das gesamte Wirtschaftsleben dieser Welt und damit für unser aller Dasein auf dieser Erde sowohl in der Gegenwart als auch in absehbarer Zukunft richtungsweisend sein wird»...
In ähnlicher Weise vergleiche Aussagen wie z. B.: «Sir G. Edwards... new engineering techniques... spring from aeronautical endeavours («Aeroplane», 18. 11. 60., S. 677);
«Flugkörperbau,... Erkenntnisse, die man besonders in den USA und in England aus den ganz neuen Anwendungsgebieten für Kunststoffe gewinnt und die auch für andere Gebiete der Technik ausstrahlen» («Flugkörper», 1/1960, S. 4);
«Glasfaserverstärktes Giessharz (Epoxy) für Hochspannungsschaltgerät, ein dynamisches Problem der Kunststofftechnik wie in der Raketentechnik... interessante Entwicklungsmöglichkeiten gegeben!» (Dr. Weigelt von Voigt und Haeffner, Fafm. in «Chem. Ind.», Okt. 1959, S. 533);
«The aircraft industry is today the largest user of epoxy formulations for non-coatings applications»... («Epoxy Resins», Market Survey, Harvard University, Juli 1959).

[70] Trotzdem dem «Kunstharzlot» vorerst einmal eine hohe elektrische Isolationsgüte eigen war, so setzten im Zuge der «Formulierarbeiten» bald auch Versuche ein, ein «Kunstharzlot» mit einer praktisch brauchbaren elektrischen Leitfähigkeit bei im Prinzip unveränderter Bindefunktion zu entwickeln; vergleiche z. B. Aufsatz: Kilduff and Benderly, «Conductive Adhesive for Electronic Applications» («Improved silver epoxy formulations have a dual purpose: mechanical as well as electrical connection between various materials»), «Electrical Manufacturing», Juni 1958, S. 148 ff. Vergleiche auch: «Modern Plastics Encyclopedia», 1958, S. 178: «...«Eccobond»... epoxy based adhesive, paste form, to produce low electrical resistance solid,... used where soldering is impractical...» «Modern Plastics», Dez. 1962, S. 206: «Conductive Epoxy Adhesive», «An electrically conductive thixotropic epoxy

based on an article by F. B. Stencel in "Aluminium Suisse", September 1956, pp. 119 ff.,: "New lightweight construction techniques for beam-riding Oerlikon anti-aircraft missile." For further developments see also the article "Light-weight construction techniques for anti-aircraft and anti-tank missiles" by H. P. Schneiter (Contraves AG, Zurich) in "Flugwehr und Technik", January/February 1964. This article gives an account of "Bonding Agent CY-175", used for the construction of pressure vessels made of steel bands and wires. This "Bonding Agent CY-175" is the product ARALDITE CY-175, a cycloaliphatic epoxy resin described by R. Stierli (loc. cit., Note [49] above). On pages 544 and 545 it is reported that the new bonding agent still exhibits high (though somewhat reduced) strength at elevated temperatures when its mechanical properties are improved by the use of reactive plasticisers so that they are once again equal to those of the classic ethoxyline resins (such as ARALDITE casting resin B – the original de Trey resin).

[66] In missile construction the nozzles and blast tubes of solid-propellant rocket motors, as well as fuel containers and pressure vessels, are made mainly by the filament winding process, particularly in combination with glass fibre. Epoxy resins are used in applications (rocket cases of Polaris, Minuteman, Atlas/Rocketdyne motors, Titan rocket engine, etc.) where maximum strength is essential. Thus an editorial entitled "Aerospace Age" in the February 1962 issue of "Modern Plastics" (New York) states: "...The filament-wound structure offers a strength-to-weight ratio that is three times more efficient than titanium.... The Polaris missile, for example, with first and second-stage motors encased in steel, had a range of 1380 nautical miles; with a filament-wound epoxy-glass second stage, the range was increased to 1725 miles; and with 2875 miles as the goal, work is progressing on winding the cases of both first and second stages."

[67] See "Modern Plastics", May 1962, pp. 110 ff. (report on the 19th Annual Conference of the Western Section of the SPI): "D. M. Hatch, Manager of Aerospace Plastics Engineering for the H. I. Thompson Fiber Glass Co., Gardena, Calif., illustrated with slides and test-firing movies his spectacular account of the NOVA-scale and larger plastic nozzles for the mammoth solid-propellant rockets now under investigation by USAF and NASA. Hatch electrified the packed meeting room on Macrostructures with details of a proposed plastic nozzle.... He also stated that filament-wound fibrous glass and epoxy solid motor cases up to 50 ft. in diameter are under design and development by the plastics industry in California."

[68] See, for example:
"Modern Plastics", New York, October 1961, article by W. B. Royall and R. W. Matlock entitled "Epoxy passes toughest test: Outer space": "...The success of the Courier 1-B satellite will be of more than passing interest for industrial users concerned with the performance of reinforced epoxies under conditions of extreme heat and cold, intense radiation, and high vacuum. This space vehicle, still in orbit more than a year after launch, relies heavily on glass-reinforced epoxy for its structure. Its continued presence in the sky makes the material a promising candidate for more earthbound applications where service environments are severe and critical."
"Flight International", London, 19 April 1962, p. 612: "US/UK Satellite. Structure: Basic structural material in the satellite is plastic bonded glass-fibre. The central body section is an epoxy-bonded monofilament-wound cylinder structure; the upper dome is bonded to an aluminium ring which mates with an aluminium ring bonded to the mid skin, an aluminium cylinder... bonded to this are eight glass-fibre stiffening ribs...."

[69] In an introduction to the first issue of the journal "Flugkörper" (February 1959), Professor Ludwig Erhard, at that time Federal German Minister for Economic Affairs (now Federal Chancellor), discussed the importance attaching to the success of a new material or technique in the aircraft and missile industry: "It is scarcely necessary to point out that the aircraft industry is the pacemaker of technological progress, which in turn, both at the present time and in the foreseeable future, will determine the economic prosperity of the world and hence the existence of all of us who live on this planet."
For similar verdicts, see:
"Aeroplane", 18 November 1960, p. 677: "Sir George Edwards... new engineering techniques... spring from aeronautical endeavours."
"Flugkörper", No. 1, 1960, p. 4: "Aircraft industry... experience which has been acquired, especially in the United States and the United Kingdom, in the use of plastics for entirely new applications and which promises to have a significant impact on other sectors of engineering."
Dr. Weigelt (Voigt und Haeffner), in "Chemische Industrie", October 1959, p. 533: "Glass fibre-reinforced epoxy casting resin for high tension switchgear, a problem of dynamic stress in both plastics technology and rocket engineering... highly interesting development potential."

adhesive ... developed by Resin Formulators Co. ... used where hot soldering is costly or impractical»...

[71] In dem kürzlich erschienenen, in seiner Aktualität auffallenden und in seiner Aussage für unzählige Fälle stellvertretenden Aufsatze aus dem Zentrallaboratorium des Brown-Boveri-Elektro-Konzerns («Technische Rundschau», Nr. 43, 11. 10. 63) fällt z. B. dem Äthoxylin-(Epoxy-)Harz als «Tränkmittel» oder «Imprägnierharz» die Aufgabe zu, in «elektrischen Isolationssystemen», z. B. «Wicklungen und Spulen» (wie solche z. B. bei Messwandlern vorhanden sind), «einzelne Leiter», – wie der Bericht lautet – «zu einem festen Verbande zu verkleben». Und der Bericht fährt fort: «Folglich ist die Klebekraft solcher Bindemittel ... ein wichtiger Qualitätsfaktor»... «Ein Mass für die Verklebungsfestigkeit gibt der Biegeversuch»...

[72] Über diese Begriffe vergleiche u.a.: «Plastics Engineering Handbook of the SPI», 3rd Edition, 1960, S. 217.
«Electrical Encapsulation» by Volk, Lefforge and Stetson, 1962, S. 3 und 27: «...Encapsulating materials are expected to provide much of the physical integrity of the finished unit; therefore a property at least as important as the physical strength of the encapsulant is its adhesion to the rather wide variety of materials occuring in electronic components»... «Thus the epoxies... have particularly good adhesion»...
«Concise Guide to Plastics» by H. R. Simonds, 1957, S. 190/191.

[73] Charles A. Harper von der Westinghouse Electric Corp. sagt in seinem Buche «Electronic Packaging with Resins», 1961, S. 2, aus: «...The processes commonly used for packaging electronic equipment are embedment, encapsulation and impregnation. The terms embedment, potting, casting, encapsulation and impregnation are widely used. Most persons in the industry use some of the terms interchangeably»...

[74] Vergleiche z. B. J. Delmonte im Handbuch «Metal-filled Plastics», Reinhold/New York, 1961, S. 117: «The encapsulation of electrical conductors. Epoxy type which upon baking after assembly weld the structure into a strong, stable unit. Large horsepower motors and generators utilize copper windings which are welded together with the aid of tapes and plastic impregnants. After assembly with the motor laminations and the frame, the entire motor or generator may be further impregnated to eliminate dead air spaces; moreover, further baking or cure welds the assembly into a permanent structure.» ... «Though we will not be diverted by the field of resin encapsulation and impregnation of electrical units, ranging from the potting of joined cable terminals in the field to the factory impregnation and encapsulation of transformers and electronic circuitry, the uniqueness of solid dielectrics and concomitant electrical conductors must be mentioned. These truly represent functional marriages of plastics and metals.» S. 204: «Through plastic binders, unorthodox distributions of metallic elements become a reality»...
Ähnlich schreibt J. O. Turner vom Lawrence Radiation Laboratory, Berkeley der University of California im Handbuch «Plastics in Nuclear Engineering», 1961, Reinhold/New York, S. 65: «...Use of epoxy resin embedment is quite common for coils of all sizes from tiny magnetic field survey coils to magnet coils with great mechanical and electrical ruggedness. As a result, it has become standard procedure where toughness, mechanical strength and accuracy are important.» S. 82/83: «Brookhaven AG Synchrotron... main magnet coils ...constructed of rectangular copper conductor (ca. 4 by 2 cm)... The assembled coil is ... served with woven fiber glass tape and hot pressed to melt and cure the epoxy adhesive film to a monolithic mass»... S. 89: «Vacuum tanks. ... When the electron synchrotron at Berkeley was built, an attempt was made to construct its vacuum tank with an unsaturated polyester resin and fiber glass. It could never be made vacuum tight, so that material was abandoned. ... The search ended with the development of a technique based on fiber glass and epoxy resin. ... When the piece went into service in the machine, it proved to have good vacuum and mechanical properties. ... Synchrotron at Cambridge: ...vacuum tank ... a unique arrangement of metal and plastic.... woven fiber glass... epoxy resin ... metal tube... gives assembly rigidity and strength. Epoxy resins and fiber glass have better vacuum and mechanical properties than other resins and fibers, and thus are generally preferred for this use.»
Von welcher Bedeutung im Sinne des Gebrauchswertes die Bindefunktion bei Spulenkörpern der Elektrotechnik ist, beleuchtet der Aufsatz von Wainwright and Harrison (English Electric) in «Electrical Review», 7. 9. 62., S. 385 (Aufsatz: «Generator Winding Insulation, Determining Faults and Weaknesses»): «There is little doubt that in many of the cases for which information has been published, the first step in the breakdown process of a stator winding is failure of the interstrand bond. Once the conductor laminations become loose, forces of a thermal, magnetic or vibrational nature cause relative movements which result in destruction of the interstrand and sometimes the interturn insulation»...
H. Meyer, Leiter des Hochspannungslaboratoriums im Dynamo-

"Epoxy Resins, Market Survey", Harvard University Press, July 1959: "The aircraft industry is today the largest user of epoxy formulations for non-coatings applications."

[70] Although one of the outstanding features of the original "plastic solder" was its high insulating power, the formulators soon began experimenting with "plastic solders" combining good conductivity with unimpaired bonding strength: see Kilduff and Benderly "Conductive adhesive for electronic applications", "Electrical, Manufacturing", June 1958, pp. 148 ff.: "Improved silver epoxy formulations have a dual purpose: mechanical as well as electrical connections between various materials." See also "Modern Plastics Encyclopedia", 1958, p. 178: "'Eccobond' ...epoxy based adhesive, paste form, to produce low electrical resistance solid... used where soldering is impractical"; "Modern Plastics", December 1962, p. 206: "Conductive epoxy adhesive – An electrically conductive thixotropic epoxy adhesive...developed by Resin Formulators Co....used where hot soldering is costly or impractical".

[71] A recent article written by the Central Laboratory of the Brown Boveri & Cie. electrical engineering concern that is of topical and authoritative value ("Technische Rundschau", No. 43, 11 October 1963) mentions that when used as "impregnating resins" the epoxies perform the task of bonding individual conductors together so as to form a rigid unit, as for example in windings and coils of the type to be found in instrument transformers. The article continues by pointing out that the bonding power of adhesives of this type is consequently an important factor determining quality, and mentions that flexural testing may be used to obtain information regarding the strength of the bond.

[72] For definitions of these terms, see:
"Plastics Engineering Handbook of the SPI", 3rd Edition, 1960, p. 217.
"Electrical Encapsulation", by Volk, Lefforge and Stetson, 1962, pp. 3 and 27: "...Encapsulating materials are expected to provide much of the physical integrity of the finished unit; therefore a property at least as important as the physical strength of the encapsulant is its adhesion to the rather wide variety of materials occurring in electronic components.... Thus the epoxies... have particularly good adhesion...."
"Concise Guide to Plastics", by H. R. Simonds, 1957, pp. 190 and 191.

[73] In his book "Electronic Packaging with Resins" (1961) Charles A. Harper of the Westinghouse Electric Corp. says (p. 2): "...The processes commonly used for packaging electronic equipment are embedment, encapsulation and impregnation. The terms embedment, potting, casting, encapsulation and impregnation are widely used. Most persons in the industry use some of the terms interchangeably...."

[74] See, for example, J. Delmonte, writing in the manual "Metal-filled Plastics", Reinhold, New York 1961, p. 117: "The encapsulation of electrical conductors. Epoxy type which upon baking after assembly weld the structure into a strong, stable unit. Large horsepower motors and generators utilize copper windings which are welded together with the aid of tapes and plastic impregnants. After assembly with the motor laminations and the frame, the entire motor or generator may be further impregnated to eliminate dead air spaces; moreover, further baking or cure welds the assembly into a permanent structure."... "Though we will not be diverted by the field of resin encapsulation and impregnation of electrical units, ranging from the potting of joined cable terminals in the field to the factory impregnation and encapsulation of transformers and electronic circuitry, the uniqueness of solid dielectrics and concomitant electrical conductors must be mentioned. These truly represent functional marriages of plastics and metals." And on p. 204: "Through plastic binders, unorthodox distributions of metallic elements become a reality...."
A similar verdict is given by J. O. Turner of the Lawrence Radiation Laboratory of the University of California at Berkeley in the manual "Plastics in Nuclear Engineering" (Reinhold, New York 1961), p. 65: "...Use of epoxy resin embedment is quite common for coils of all sizes from tiny magnetic field survey coils to magnet coils with great mechanical and electrical ruggedness. As a result, it has become standard procedure where toughness, mechanical strength and accuracy are important." And on pages 82 and 83: "Brookhaven AG Synchrotron...main magnet coils...constructed of rectangular copper conductor (approx. 4 by 2 cm).... The assembled coil is...served with woven fiber glass tape and hot pressed to melt and cure the epoxy adhesive film to a monolithic mass...." Page 89: "Vacuum tanks.... When the electron synchrotron at Berkeley was built, an attempt was made to construct its vacuum tank with an unsaturated polyester resin and fiber glass. It could never be made vacuum tight, so that material was abandoned.... The search ended with the development of a technique based on fiber glass and epoxy resin.... When the piece went into service in the machine, it proved to have good vacuum and mechanical properties.... Synchrotron at Cambridge: vacuum tank...a unique arrangement of metal and plastic...woven fibre glass...epoxy resin...metal tube...gives assembly rigidity and

werk der Siemens-Schuckert-Werke AG, Berlin-Siemensstadt, berichtete dazu («ETZ-A, Bd. 80, H. 20, 11. 10. 59., S. 719 ff.), dass das ausgehärtete synthetische Imprägnierharz auch bei Betriebstemperatur einen festen Zusammenhalt der Glimmerbänder gewähre und dass die kunstharzgebundene Isolierung für Maschinen mit ständigem Lastwechsel Vorteile biete. Die Mikafolium-Isolierung mit Kunstharzverklebung mit Hilfe der Epoxydharzimprägnierung werde besonders dort angewendet, wo hohe Festigkeitswerte oder hohe Überlastbarkeit angestrebt werde.

[75] Vergleiche dazu: F. K. Trietsch im Aufsatz «Glasgewebeschichtstoffe und ihre Verwendungsmöglichkeiten» in «Konstruktion», Heft 11, 1954, S. 433/434.
Guy Lively, Material and Process Engineer der Douglas Aircraft Comp. in Santa Monica, in einem Vortrag über «Laminieren» am S. P. E.-Kongress in Los Angeles, Nov. 1956: «...The exceptional interlaminar bond strength of the epoxy resins, which produces greater values over the entire range of physical properties, has been recognized for many years...»...Parallel dazu äusserte sich H. D. Boggs, Generalmanager der Fibercast Corporation, in seinem Vortrage «Long term strength of reinforced Epoxy pipe»: «...The end product is a resultant of thousands of minute glue lines subject to all the laws of surface chemistry. ...We have to talk in terms of a glue industry»...
Handbuch «Glass-reinforced Plastics», Ph. Morgan, London 1955, S. 85: «...Properties of epoxide/glass laminates. ...Flexural strength...epoxide/glass laminates...higher than that of polyester ...higher tensile strength... ...these improvements are probably largely due to the better interlaminar adhesion...best illustrated by peel strength... Epoxides...higher wet strength than polyester attributable ...to better bonding to the glass...»...
Bericht in «technica», Basel, 12. 5. 61 (Nr. 10), Kapitel «Werkstoffe»: «Brennstofftanks aus Glasfasern», «...Boeing Airplane Comp. ...Glasfasertanks für die Ausrüstung von Raketen... Faden um eine Form gewickelt ,dann werden die Fäden durch Kunstharz miteinander verklebt...».
Im we gezeichneten Beitrag «Glasfaserverstärkte Kunststoffe im Wickelverfahren hergestellt» der «technica» Nr. 6/1964 vom 13. 3. 64, Seite 387, wird u. a. die 18. Tagung der Fachgruppe für verstärkte Kunststoffe in der amerikanischen «Society of the Plastics Industry (SPI)» ausgewertet. Das Wickelverfahren sei danach die Methode, die den grössten Kapazitätszuwachs aufweise. «Im Wickelverfahren wird Verstärkungsmaterial aus endlosen Glasfäden verarbeitet, um die grösste Nutzanwendung der Glasfaser-Steifigkeit zu erreichen. Glasseidenfäden oder Glasfaserstränge (Rovings) werden abgespult und durch ein Kunstharz-Bad geführt. Die mit Epoxy- oder Polyesterharz getränkten Glasfasern werden auf eine drehende Form geleitet, die den Umfang des gewünschten Endprodukts hat.»
K. H. Boller, Forest Prod. Lab., U.S. Dept. of Agriculture, Madison Wisc. im Aufsatz «Fatigue characteristics of RP laminates subjected to axial loading» (The work reported was sponsored by Air Force Materials Laboratory, Research and Technology Division, USAF), in «Mod. Plast.», Juni 1964, S. 145 ff. Beachtenswert sind im vorliegenden Zusammenhange die Fig. 6 und 12 (S. 146/148), welche das günstige Ermüdungsverhalten von Aethoxylin-(Epoxy-)Harz gebundenen Glasfaserlaminaten im Vergleich zu andern Kunstharzen hervorheben.
P. Zoller (Ciba Basel) im Aufsatz «Das Glaswickelverfahren» («Technische Rundschau», 12. 5. 64) schreibt, dass die Epoxydharze heute für dieses Verfahren beinahe ohne Ausnahme verwendet werden, weil sie die beiden Grundbedingungen, nämlich das «Zusammenhalten der einzelnen Glasfasern» und das «Verteilen der auftretenden Kräfte auf die tragenden Fäden», erfüllen.
Dr. M. Hagedorn in seinem Bericht über die 15. Tagung der Reinforced Plastics Division der SPI, 2.–4. Februar 1960, in Chicago, in «Kunststoffe», Aug./Nov. 1960, S. 549/550: «...Epoxyharze mit Füllungen aus kleinen Hohlkügelchen aus anorganischem und organischem Material wurden als neuer Werkstoff als «Epoxyharz mittlerer Dichte» eingeführt ...ungehärtet eingefüllt wirkt es gleichzeitig als sein eigener Klebstoff». S. 444: «...Die zum Aufbrechen der Faserbündel führenden Kräfte können durch Verwendung von Epoxyharzen und von geeigneten Füllstoffen vermindert werden». Und 1961 referierte Hagedorn am gleichen Orte («Kunststoffe», November 1961, S. 711): «...Man sollte Laminate nicht wie bisher üblich als von Harz umhüllte Glasfäden, sondern besser als Glasoberflächen begrenztes Harz auffassen, welches durch den Schrumpfvorgang so lange unter hohe Zugspannungen gesetzt wurde, als es sich nicht von der unnachgiebigen Glaswand lösen kann»...
J. Martel, chef du Service des Etudes techniques et économiques à l'Etablissement Fibre et Mica de la Compagnie Electro-Mécanique., Lyon, im Aufsatz «Propriétés et applications des stratifiés utilisés dans l'industrie» («Ind. des Plast. Mod.», Dez. 1964), S. 121: «Historique et Evolution. ... Avec les exigences de plus en plus sévères sur les plans électrique et mécanique associés, le tissu de verre époxyde a pris un important essor dès 1953 grâce à une excellente liai-

strength. Epoxy resins and fiber glass have better vacuum and mechanical properties than other resins and fibers, and thus are generally preferred for this use."
The importance of the bonding function in determining the ultimate performance characteristics of the finished coil is effectively illustrated by Wainwright and Harrison (English Electric) in their article "Generator winding insulation: determining faults and weaknesses" ("Electrical Review", 7 September 1962), p. 385: "There is little doubt that in many of the cases for which information has been published, the first step in the breakdown process of a stator winding is failure of the interstrand bond. Once the conductor laminations become loose, forces of a thermal, magnetic or vibrational nature cause relative movements which result in destruction of the interstrand and sometimes the interturn insulation...."
H. Meyer, head of the high tension laboratory in the dynamo plant of the Siemens-Schuckert-Werke AG, Berlin-Siemensstadt, reports ("ETZ-A", Vol. 80, No. 20, 11 October 1959, pp. 719 ff.) that the cured synthetic impregnating resin provides firm bonding of the mica tapes at service temperatures and that resin-bonded insulation affords special advantages for machines subject to constant load reversal. Resin-bonded micafolium insulation employing the technique of impregnation with epoxy resins is used particularly in cases where high strength or overload capacity are desired.

[75] See:
F. K. Trietsch, "Glass fabric laminates and their application potential" (loc. cit., Note [45] above).
Guy Lively, Materials and Process Engineer of Douglas Aircraft Company, Santa Monica, Calif., in a paper on "Laminating" read to the S.P.E. Congress held at Los Angeles in November 1956: "...The exceptional interlaminar bond strength of the epoxy resins, which produces greater values over the entire range of physical properties, has been recognized for many years...." A similar opinion was expressed by H. D. Boggs, General Manager of the Fibercast Corporation, in his paper entitled "Long term strength of reinforced epoxy pipe": "...The end product is a resultant of thousands of minute glue lines subject to all the laws of surface chemistry.... We have to talk in terms of a glue industry...."
"Glass-reinforced Plastics", Ph. Morgan, London 1955, p. 85: "Properties of epoxide/glass laminates.... Flexural strength...of epoxide/glass laminates...higher than that of polyester.... Higher tensile strength.... These improvements are probably largely due to the better interlaminar adhesion...best illustrated by peel strength.... Epoxides...higher wet strength than polyester attributable...to better bonding to the glass...."
Short item in "technica", Basle, No. 10 (12 May 1961), section headed "Materials": "Glass fibre fuel tanks": "Boeing Aircraft Comp.... Glass fibre tanks as rocket equipment.... Glass filament wound round a mandrel and then bonded together with synthetic resin...."
An article signed "we" and entitled "Glass fibre-reinforced plastics made by filament winding process" in "technica", No. 6 (13 March 1964), p. 387, gives a report of the 18th Conference of the Reinforced Plastics Section of the U.S. Society of the Plastics Industry (SPI). The filament winding technique is stated to be undergoing the greatest capacity increase. "The filament winding process utilises reinforcing material consisting of continuous glass filament in order to exploit to the full the rigidity of glass fibre. Continuous glass filament or glass rovings are wound off a reel and passed through a bath of synthetic resin. The glass fibre, now impregnated with epoxy or polyester resin, is wound round a turning mandrel with the same shape as the desired final article".
K. H. Boller of the Forest Products Laboratory, U.S. Dept. of Agriculture, Madison, Wisc.: "Fatigue characteristics of RP laminates subjected to axial loading" (sponsored by Air Force Materials Laboratory, Research and Technology Division, USAF), in "Modern Plastics", June 1964, pp. 145 ff. In this particular context special interest attaches to Figs. 6 and 12 (pp. 146 and 148), which emphasise the better fatigue behaviour of epoxy resin-bonded glass fibre laminates when compared with laminates made with other synthetic resins.
P. Zoller (CIBA Limited, Basle) writes in his article "The glass filament winding process", "Technische Rundschau", 12 May 1964, that the process today makes use almost exclusively of epoxy resins, since the latter satisfy the two fundamental requirements, viz. bonding together of the individual glass fibres and distribution of the resulting forces over the supporting fibres.
Dr. M. Hagedorn, reporting in "Kunststoffe" (August/November 1960) on the 15th Annual Technical and Management Conference of the Reinforced Plastics Division of the SPI held at Chicago from 2 to 4 February 1960, writes (pp. 549 and 550): "Epoxy resins with fillers consisting of small hollow spheres of inorganic or organic material were introduced as 'medium-density epoxy resins'; when put in place in the uncured state the new material acts as its own bonding agent." And on page 444: "The forces responsible for rupturing the fibre bundles may be reduced by the use of epoxy

son naturelle de la résine époxyde et de la fibre de verre.... L'excellente résistance au cisaillement du tissu de verre époxyde...»

[76] Die spontane, technische Entwicklung der glasfaserverstärkten Kunststoffe (GFK) setzte 1941 mit dem Eintritt Amerikas in den II. Weltkrieg ein. (Vergl. z.B. H. Hagen, «Glasfaserverstärkte Kunststoffe», Berlin, 1961, S. 1 und «Polymer Processes», Interscience-Verlag, New York, 1956 (Vol. X der Reihe «High Polymers»), S. 761/762.)

[77] Vergleiche dazu: D. J. Duffin im Handbuch «Laminated Plastics», New York, 1958. S, 81: «Plastic Tooling: One of the biggest industrial applications of low pressure laminates is in the field of plastic tooling – the manufacture of tools from epoxy-impregnated glass fiber, in the form of metal-forming tools (stretch dies, draw dies, rubber press forms, drop hammer dies); holding fixtures, gauges, foundry tools ... prototype production tools ... and plastic forming tools.»
Handbuch «Plastics Tooling» (M. W. Riley, New York 1961: S. 31: «The strength of reinforced plastics is dependent not only on the strength of the bond between resin and fiber. Also the overall load bearing ability of such a composite is dependent on the highly complex stress and strain distributions within the mass.» (Vergl. dazu auch Anm. 75). Guy Livelys Aussage: «The exceptional interlaminar bond strength of the epoxy resins, which produces greater values over the entire range of physical properties, has been recognized for many years».) S. 78: «...6. Reinforcements for Tooling. Plastics resins themselves are relatively low strength, low modulus materials. Fibrous reinforcements contribute the primary structure strength. Consequently, the strength characteristics of a laminate are highly dependent ... on the degree of adhesion between the resin and the glass.» S. 47: «Epoxy laminates reinforced with glass cloth can provide the highest strengths of any low pressure reinforced plastic material»...

[78] Vergleiche z.B.: Bericht der Technischen Hochschule in Karlsruhe in «Kunststoffe», November 1962, «Mechanische Beanspruchungen der Kunststoffe im Massivbau», S. 662: Abschnitt «Füllung von kraftübertragenden Fugen»: «Im Metallbau hat sich bereits die Verwendung von Giessharzen (gemeint sind Epoxygiessharze) für die kraftschlüssige Verbindung von Fugenteilen bewährt»... «Als Kleber werden meist ungefüllte Giessharze verwendet. Sollen Fertigteile aus Beton durch Klebung zu einer Konstruktion verbunden werden, so sind in der Klebfuge unvermeidliche Herstellungsungenauigkeiten auszugleichen. ...Fugendicke 2 bis 15 mm. Zur Füllung dieser Fugenweite sind Giessharze nur in Verbindung mit einem zweckentsprechenden Zuschlag, meist Kiessand geeignet»... «Für die Zusammensetzung von Estrichen und Ausgleichsschichten (Flickmörteln) gelten gleichfalls die für kraftübertragende Fugen angeführten Gesichtspunkte»...
«Neue Zürcher Zeitung», Technische Beilage, 23. 8. 62, Aufsatz: Die Verwendung von Epoxyharzen in der amerikanischen Bauindustrie»: «...Im folgenden sollen nun die Epoxyharze besprochen werden, deren Verwendungsmöglichkeiten im amerikanischen Baugewerbe in den letzten Jahren, der erstaunlichen chemischen und mechanischen Eigenschaften wegen, sehr stark zugenommen hat. ...Äusserst stark ist die Adhäsion zu andern Baumaterialien. ...Epoxyprodukte widerstehen Vibrationen. ...In Amerika werden von gewissen Regierungsstellen eigene Epoxykompositionen vorgeschrieben, die für alle Arbeiten, die diesen Organen unterstehen, verwendet werden müssen»...
Aufsatz Pollet, Jacotot und Peignard «La rentabilité du béton à liant plastique» in «Industrie des Plastiques Modernes», Mai 1964. Es wird über Versuche berichtet, bei welchen «gravillons» (10–40 mm Ø) und «sable» (0,4–10 mm Partikeldurchmesser) mittels ARALDIT M, «qui sert de colle» – wie die Verfasser festhalten – verklebt werden zu einem Beton, welcher «a bien des chances de devenir le béton de demain.»
Bericht «Forschung in der Industrie» (SHELL) in «Neue Zürcher Zeitung» («Technik»), 25. 11. 64: SHELL verfestigt den Lockersand (Sandkörner) in ihren Erdölbohrstellen mittels Äthoxylin-(Epoxy-)Harz zu solcher Festigkeit, wie dies bis anhin mit keinem andern organischen oder anorganischen Bindemittel gelang (siehe Bild 18).
«Yearbook of Science and Technology» (1962 Review/1963 Preview, McGraw Hill Book Comp. Inc., New York, Chikago, San Francisco, Dallas, Toronto, London), S. 206, Kapitel «Construction Engineering»: «...High among the significant advances in the field of construction materials in 1962 were the continued development and widespread acceptance of epoxies. When properly compounded and applied, *epoxies will provide a virtually unbreakable bond between almost any materials, similar or dissimilar.* They have done *more* than provide a new product; they actually make *possible completely new methods of construction.*»... (Auch hier gilt wie in praktisch allen andern Applikationssektoren der Technik die andauernde und weltweite Aktualität der die Äthoxylin-(Epoxy-)Harze zentral kennzeichnenden Bindefunktion!).

[79] Vergleiche dazu: Aufsatz «Safe Roads that last» in «Modern Plastics», Februar 1960, S. 102ff.: «Long service life and improved

resins and of suitable fillers." Writing in the same journal ("Kunststoffe", November 1961) in 1961, Hagedorn reported (p. 711): "Laminates should be regarded not as glass fibres surrounded by a resin envelope – as has hitherto been the case – but rather as resin that is bounded by glass surfaces and that is subjected to high tensile stresses by the process of shrinkage because it is unable to detach itself from the unyielding surface of the glass fibres."
J. Martel, Chef du Service des Etudes techniques et économiques à l'Etablissement Fibre et Mica de la Compagnie Electro-Mecanique, Lyons, in his article "Propriétés et applications des stratifiés utilisés dans l'industrie" ("Ind. des Plast. Mod.", Dec. 1964), p. 121: "History and Development.... Because of increasingly stringent requirements from both the electrical and mechanical points of view, glass fibre-epoxy laminates have assumed considerable importance in view of the excellent natural bond between the epoxy resin and the glass fibre.... The outstanding shear strength of the glass fibre-epoxy laminate...."

[76] The development of glass fibre-reinforced plastics began spontaneously in 1941 with the entry of the United States into World War II. See, for example, H. Hagen, "Glasfaserverstärkte Kunststoffe", Berlin 1961, page 1, and "Polymer Processes", Interscience Publ., New York 1956, pp. 761 and 762 (Vol. X of the series "High Polymers").

[77] See:
D. J. Duffin in "Laminated Plastics", New York 1958, p. 81: "Plastic Tooling: One of the biggest industrial applications of low pressure laminates is in the field of plastic tooling – the manufacture of tools from epoxy-impregnated glass fiber, in the form of metal-forming tools (stretch dies, draw dies, rubber press forms, drop hammer dies); holding fixtures, gauges, foundry tools ... prototype production tools ... and plastic forming tools."
"Plastics Tooling" (M. W. Riley, New York 1961, p. 31): "The strength of reinforced plastics is dependent not only on the strength of the bond between resin and fiber. Also the overall load bearing ability of such a composite is dependent on the highly complex stress and strain distributions within the mass." (Cf. Note [75] above: Guy Lively's statement that "The exceptional interlaminar bond strength of the epoxy resins, which produces greater values over the entire range of physical properties, has been recognized for many years.") And on page 78: "6. Reinforcements for Tooling. Plastics resins themselves are relatively low strength, low modulus materials. Fibrous reinforcements contribute the primary structure strength. Consequently, the strength characteristics of a laminate are highly dependent ... on the degree of adhesion between the resin and the glass." Page 47: "Epoxy laminates reinforced with glass cloth can provide the highest strengths of any low pressure reinforced plastic material."

[78] See, for example:
Report by the Karlsruhe Technische Hochschule in "Kunststoffe", November 1962, entitled "Mechanical stresses in plastics in solid structures", p. 662, section headed "Filling of load-transmitting joints": "The use of casting resins (the reference is to epoxies) for creating a high strength bond between parts to be joined has provided satisfactory results in metal construction work. Casting resins containing no fillers are generally used. If finished components made of concrete are to be united into a structure by bonding, any inaccuracies, which are unavoidable, must first be evened out. Gaps of 2 to 15 mm: For gaps of this size casting resins should be used only in combination with an appropriate additive, generally sand. The rules listed for load-transmitting joints also apply to the laying of floors and compensating layers."
"Neue Zürcher Zeitung", Technical supplement, 23 August 1962, article entitled "The use of epoxy resins in the American building industry": "The following is an account of the epoxies, a class of resins whose use by the American building trade has increased sharply in recent years because of their outstandingly good chemical and mechanical properties. Their adhesion to other building materials is exceptionally strong, and they withstand vibration. In the United States certain government agencies prescribe their own epoxy formulations which have to be used in all building work for which they are responsible."
Pollet, Jacotot, and Peignard, "The economics of plastic-bound concrete", in "Industrie des Plastiques Modernes", May 1964, a report on trials carried out to investigate the bonding of "gravel" (10 to 40 mm diameter) and "sand" (0.4 to 10 mm particle diameter) with ARALDITE M – which, as the authors point out, acts as the binding agent. The resulting concrete "has every chance of becoming the concrete of the future" ("a bien des chances de devenir le béton de demain").
"Research in industry", "Neue Zürcher Zeitung", 25 November 1964: it is reported that Shell use epoxy resins to bind loose sand grains in oil wells and that the strength obtained is greater than that hitherto possible with any other organic or inorganic binding agent (see Fig. 18).
"McGraw-Hill Yearbook of Science and Technology 1963", McGraw-Hill Book Co. Inc., New York, etc., p. 206, entry on "Construction Engineering": "High among the significant ad-

safety are the major contributions epoxies bring to road constructions.... The typical resin system is a two-component material consisting of the epoxy resin and the catalyst diethylenetriamine ... mixed with sand to form a thick «mortar» which is excellent for repairing.
F. K. Trietsch anlässlich der Vortragsreihe «Kunststoff-Verwendung im Strassenbau» im Haus der Technik in Essen am 25. Januar 1962 («Kunststoffe», Mai 1962, S. 271): «Die Anwendungen im Strassenbau beruhen vorwiegend auf ihrer Klebwirkung. Voraussetzung für eine zuverlässige und hohe Haftung auf Beton sind gesunde und saubere Oberflächen. ...Die Hauptanwendungen der Epoxydharzmörtel als Betonkleber sind: Verkleben von Altbeton mit Altbeton, Haftbrücke von Altbeton zu Neubeton, Flicken von Rissen, Aufkleben von Fahrbahnmarkierungen»...)
Rolf Müller in «Technische Rundschau» (loc. cit. Anm. 3), S. 13: «...Alle Anwendungen der Epoxyharze im Strassenbau basieren vorwiegend auf ihrer Klebwirkung. Mit ihnen lassen sich Beton mit Beton verkleben... Das Kleben von Beton mit Epoxydharzen beginnt im gesamten Bauwesen als Verbindungsverfahren ähnliche Bedeutung zu gewinnen wie die Verklebung von Metallen in vielen Industriezweigen. Für die Verklebung von Altbeton mit Altbeton und als Haftbrücke von Altbeton zu Neubeton sind die Epoxyharze hervorragend geeignet. Ihr ungewöhnliches Haftvermögen, das auch für die hohe Festigkeit der Harz-Füllstoff-Mischungen entscheidend ist, führt bei guten Verklebungen stets zum Bruch im Beton»...

[80] Vergleiche Vortrag von Dr.-Ing. Karl Mienes an der 9. Deutschen Kunststofftagung am 12. 4. 62 «Die Kunststoffdekade 1960/1970» (publiziert bei Hanser, München, 1961, als Sonderdruck): «Das Einbinden der eigentlichen Fahrbahn mit Kunststoffen bietet auch in anderer Weise Vorteile für Haltbarkeit und Sicherheit im zunehmenden Strassenverkehr. Epoxyharzbindung verhindert z. B. das beim Teerbelag unvermeidliche Absetzen der keramischen Zuschlagstoffe und gewährleistet dauerhafte Griffigkeit der Deckschicht» ... Vergleiche auch: «Modern Plastics», Nov. 1962, S. 246: «...epoxy resin-system employed to bond the terazzo directly to a concrete slab... The purpose of the epoxy is to bond»...

[81] Assistant Director of Research in seinem Aufsatz «Structural Interplay: Design and Materials», Zeitschrift «Aero Space Engineering» Aug. 1959, S. 42.

[82] Vergleiche George Epstein, Research Laboratory, Aeronutronic Division of Ford Motor Company, Newport Beach, Calif. anlässlich des «Symposium on Adhesives for Structural Applications», held at Piccatinny Arsenal, Dover, New Jersey, Sept. 27–28, 1961 (Sonderheft des «Journal of Applied Polymer Science», Vol. VI, März-April 1962) in seinem Aufsatze «Potential of Adhesives for the Future»: «... Considering its recent origin, advances in adhesive technology have made remarkable progress – and augur for the future potential of adhesives. ...Of particular advantage ... is the fact that adhesives lend themselves ideally to joining of dissimilar materials – a very significant factor in the rapid-growing field of composites. A composite material system may be defined as a combination of two or more materials into a unit, the component materials being arranged so as to confer upon the resulting unit the desired properties of each material employed ... epoxy base adhesives have helped to further the potential of adhesives.»...
In paralleler Weise stellten kürzlich K. H. Pohl und A. T. Spencer (Bell Telephone Lab. Inc. Holmdel and Murray Hill, N.J.) in ihrem Aufsatze «New Structural Laminate: Polyethylene core, aluminum skins» («Mod. Plast.», März 1964, S. 121) fest: «...Most materials encountered in engineering today may be described as homogeneous, that is, consisting of a uniform material over the entire cross-section. With the exception of such widely used building materials as reinforced concrete, non-homogeneous materials generally have not found wide use in engineering. This may be largely attributed to the *lack of reliable methods for bonding* the elements of these composite materials or their high cost»... Die Bresche schlugen dann bekanntlich die Äthoxylinharze!
Über die weiteren Ausstrahlungen des modernen Begriffes «Composite material» bis zurück auf das eigentliche Gebiet der Metallurgie vergleiche z. B. den Aufsatz «Fiber-Reinforced Metals» in «Scientific American», New York, Febr. 1965.

[83] Epstein (loc. cit. Anm. 82) zuerst bei Northrop, nachher Head of Research and Development Group, Structural Plastics Dept., Aerojet-General Corp., Azusa, Calif., gleichzeitig Instructor in Resins and Adhesives for Engineering Extension, University of California, Los Angeles) schrieb 1954 in seinem Buche «Adhesive Bonding» auf S. 10: «...«Encapsulating» and «Potting» of electrical components, which can be regarded as a special application of adhesives.»...
In gleicher Weise definierte John Delmonte anlässlich der Konferenz «Metal-to-Metal Adhesives for the Assembly of Aircraft», welche an der University of California mit Unterstützung der Aircraft Association of America Inc. durchgeführt worden war (Sept.

vances in the field of construction materials in 1962 were the continued development and widespread acceptance of epoxies. When properly compounded and applied, *epoxies will provide a virtually unbreakable bond between almost any materials, similar or dissimilar*. They have done more than provide a new product; *they actually make possible completely new methods of construction*". (Here again, as in virtually every other field in which they are used, it is the characteristic bonding function of the epoxies that has ensured their continued world-wide acceptance.)

[79] See:
"Safe roads that last", in "Modern Plastics", February 1960, pp. 102 ff.: "Long service life and improved safety are the major contributions epoxies bring to road construction. The typical resin system is a two-component material consisting of the epoxy resin and the catalyst diethylenetriamine...mixed with sand to form a thick 'mortar' which is excellent for repairing...."
F. K. Trietsch, contribution to the series of lectures on "The use of plastics in highway engineering", Haus der Technik, Essen, 25 January 1962 ("Kunststoffe", May 1962, p. 271): "Applications in road building depend mainly on the adhesive properties of the resins. An essential requirement for reliable and solid bonding to concrete is sound and clean surfaces. The principal uses of epoxy resin mortars as concrete-bonding agents are the bonding of old concrete to old concrete and of old concrete to new concrete, the repair of cracks, the application of surface markings...."
Rolf Müller in "Technische Rundschau" (loc. cit., see Note [3] above), p. 13: "The applications of epoxy resins in road engineering depend mainly on their adhesive properties. They may be used for bonding concrete to concrete. The bonding of concrete with epoxy resins is rapidly assuming the same importance in the entire field of construction engineering as the bonding of metals in other sectors of industry. The epoxies are ideally suited for the bonding of old concrete to old concrete and of old concrete to new concrete. Their outstanding adhesive power, which is also responsible for the high strength of the resin-filler mixture, is such that when a break occurs it is in the concrete rather than in the bonding agent."

[80] Compare the paper read by Dr. Karl Mienes to the 9th German Plastics Congress on 12 April 1962, "The Plastics Decade 1960–1970"; published separately by Hanser, Munich 1961: "The use of plastics for binding the actual surface of the road also affords other advantages, particularly durability and safety with increasing traffic volume. Thus epoxy binding prevents the settling of ceramic aggregate material that is otherwise inevitable, and ensures a lasting grip". – See also "Modern Plastics", November 1962, p. 246: "...epoxy resin system employed to bond the terrazzo directly to a concrete slab.... The purpose of the epoxy is to bond....".

[81] Assistant Director of Research, in his article "Structural interplay: Design and materials" in "Aero Space Engineering", August 1959, p. 42.

[82] Compare the paper read by George Epstein, Research Laboratory, Aeronutronic Division, Ford Motor Company, Newport Beach, Calif., to the Symposium on Adhesives for Structural Applications held at Piccatinny Arsenal, Dover, N.J., from 27 to 28 September 1961 (published as supplement to "Journal of Applied Polymer Science, Vol. VI, March/April 1962), entitled "Potential of adhesives for the future": "...Considering its recent origin, advances in adhesives technology have made remarkable progress – and augur for the future potential of adhesives.... Of particular advantage... is the fact that adhesives lend themselves ideally to the joining of dissimilar materials – a very significant factor in the rapid-growing field of composites. A composite material system may be defined as a combination of two or more materials into a unit, the component materials being arranged so as to confer upon the resulting unit the desired properties of each material employed...epoxy-base adhesives have helped to further the potential of adhesives...."
Similar opinions were expressed recently by K. H. Pohl and A. T. Spencer (Bell Telephone Laboratories Inc., Holmdel and Murray Hill, N.J.) in their article "New structural laminate: Polyethylene core, aluminium skins" ("Modern Plastics", March 1964, p. 121): "...Most materials encountered in engineering today may be described as homogeneous, that is, consisting of a uniform material over the entire cross-section. With the exception of such widely used building materials as reinforced concrete, non-homogeneous materials generally have not found wide use in engineering. This may be largely attributed to the *lack of reliable methods for bonding* the elements of these composite materials or their high cost...." This lack, of course, has been more than made good by the epoxies.
Regarding the further implications of the recently formulated term "composite material" in the actual field of metallurgy itself, cf. the article "Fiber-Reinforced Metals" in "Scientific American" New York, Febr. 1965.

[83] Epstein (loc. cit., see Note [82] above), first with Northrop, later Head of Research and Development Group, Structural Plastics

1954), in seinem Vortrage «Factors influencing Bonding Techniques» im Kapitel «Adhesives for Potting»: ...«Newer functions which adhesives are called upon to fulfill will of necessity employ new techniques. For example, the widespread use of potting or encapsulation compounds to seal and protect electronic components ...Good adhesion to container walls as well as to components is necessary to exclude damaging moisture. This particular application requires more than passing concern ... the potting material shrinks *(which after all is the adhesive in a larger mass)* ... In these and other ways, the functions of adhesives may necessitate new techniques in order to make full use of their desirable characteristics ... good adhesion of epoxies to metals ... attractive for the encapsulation of electronic components.»
«Zeitschrift für Metallkunde», 47 (1956), Heft 7 (K. Meyerhans): «Metallverbindungen mit Kunstharzen»: «...Nicht immer liegen für einen Verbindungsprozess dünne Fugen vor, sondern es sind sehr oft aus konstruktiven Überlegungen (in der Elektroindustrie z. B. aus isolierungstechnischen Gründen) Zwischenräume bis zu beachtlicher Stärke mit Bindemittel auszufüllen. Dieses Verbinden wird oft zwangsläufig zu einem Einbetten, Eingiessen oder Umhüllen. Dank der angeführten Eigenschaften lassen sich im besonderen Äthoxylinharze verwenden, um gleiche oder verschiedenartige metallische oder nichtmetallische Werkstoffe ohne Rücksicht auf ihre Struktur, Form oder gegenseitige räumliche Lage, in einem oder mehreren Arbeitsgängen zu einem mechanisch festen ... elektrisch hochwertigen Verbande vereinigen».
Zeitschrift «Kunststoffe», Dezember 1951 (K. Meyerhans), S. 6/7: «...Giessharz (Verarbeitung und Härtung): Wenn ARALDIT in grösserer Schichtdicke angewendet werden soll, so kommt das ARALDIT-Bindemittel Typ I infolge seiner relativ hohen Viskosität nicht mehr in Frage. Für solche Zwecke steht ARALDIT-Giessharz B zur Verfügung. ...Irgendwelche Apparateteile nicht zu verbinden, sondern zur gleichen Zeit auch umhüllen lassen».
Inserat eines «Formulator», der FURANE PLASTICS Inc., Los Angeles, in «Modern Plastics Encyclopedia for 1962», S. 366: «The Epoxy of a thousand commercial uses: The only all-purpose bonding material, provides super-strength bonds – to practically any surface...: Electrical (Encapsulating, potting, dip coating of connectors, cable terminations, etc.); Building Trades (Repairing cracks..., impregnating..., grouting..., repairing and fabricating...); Marine (Glass cloth lamination, ...lining, filling and sealing..., bonding..., potting...); Automotive (Smoothing and filling..., general metal repair); Tooling (Casting..., gelcoating..., filling...)».

[84] In einer redaktionellen Übersicht («The Plastiscope») der Mai-Ausgabe 1964 von «Modern Plastics» schreibt im Abschnitt «Epoxy resins continue growth pattern» R. L. van Boskirk treffend: «...And new markets will continue to be developed and won because epoxies have almost unbelievable bonding and electrical properties and are extremely useful in thousands of relatively small applications.» Und in einem anschliessenden Aufsatz «Epoxy makers expand capacity for specialty needs» der gleichen Übersicht («The Plastiscope») der Juli-Nr. 1964 wird die Schlussfolgerung gezogen: «...The only material with which epoxies could be competitive are polyesters; and any price drop short of reaching the polyester level of the low twenties would have little effect on total epoxy sales. This fact is well recognized by all companies in the field; and they all aim *their sales effort at those areas where the exceptional properties of epoxy resins make them the only suitable choice.*» (Vergleiche auch Anm. 4 und Anm. 63.)

[85] Vergleiche L. J. Broutman and F. J. McGarry, respectively Research Assistant and Associate Professor of Materials, Departmen of Civil Engineering, Massachusetts Institute of Technology, Cambridge, Mass. (MIT) im Aufsatz «Glass-resin joint strength studies», «Modern Plastics», September 1962, S. 161 ff. (The research described in this report was performed under Contract No. AF33 (616)-6280, Wright Air Development Division. It was executed in the MIT Plastics Research Laboratory which is supported by the Manufacturing Chemist's Association Inc.): Optische (Photoelastische) Untersuchungen an Kunstharz verbundenen Glasfasern führten zum Ergebnis, dass «Epoxy resins have higher bond strengths than do the polyester resins»...

[86] Vergleiche: A. Puck, Deutsches Kunststoffinstitut Darmstadt, in: a) «Kunststoffe», März 1963, Aufsatz «Dimensionierung tragender Leichtbaukonstruktionen aus GFK»: Der Autor bezeichnet bei der modellmässigen Betrachtung des Verbundstoffes aus Harz und Glasfasern denselben als eine kraftübertragende Packung von Glasstäben, die durch ein Bindemittel verklebt sind. Als Kunstharz wird Äthoxylin-(Epoxy-)Harz verwendet. – b) «Kunststoffe», Okt. 1963, Aufsatz: «Glasfaser/Kunststoff in hochbeanspruchten Leichtbaukonstruktionen»: Erfahrungen mit dem Verbundstoff (Äthoxylin-(Epoxy-)Harz gebunden) in der Praxis.
H. Wurtinger, Deutsches Kunststoffinstitut, Darmstadt in «Kunststoffe», April 1963, Aufsatz: «Bau und Erprobung einer Segelflugtragfläche in Glasfaserkunststoff-Bauweise.»

Dept., Aerojet General Corp., Azusa, Calif., and at the same time Instructor in Resins and Adhesives for the Engineering Extension, University of California, Los Angeles, refers in his book "Adhesive Bonding" (1954), p. 10, to the "...'Encapsulating' and 'potting' of electrical components, which can be regarded as a special application of adhesives...."
John Delmonte spoke on the same topic at the conference on "Metal-to-Metal Adhesives for the Assembly of Aircraft" organised by the University of California with the support of the Aircraft Association of America Inc. (September 1954); in his paper entitled "Factors influencing bonding techniques", he said (in the section devoted to "Adhesives for potting"): "...Newer functions which adhesives are called upon to fulfill will of necessity employ new techniques. For example, the widespread use of potting or encapsulation compounds to seal and protect electronic components.... Good adhesion to container walls as well as to components is necessary to exclude damaging moisture. This particular application requires more than passing concern...the potting material shrinks *(which after all is the adhesive in a larger mass)*.... In these and other ways, the functions of adhesives may necessitate new techniques in order to make full use of their desirable characteristics...good adhesion of epoxies to metals... attractive for the encapsulation of electronic components...."
"Zeitschrift für Metallkunde", Vol. 47 (1956), No. 7, article by K. Meyerhans, "Bonding of metals with synthetic resins": "The gap between the surfaces to be bonded together is not always small; in fact it frequently happens that gaps of considerable dimensions have to be filled with adhesive, as for example in electrical engineering, where insulation is also a factor to be considered. In such cases bonding may embrace embedding, encapsulation, or encasing. In view of their specific properties, ethoxyline resins are highly suitable for uniting similar or dissimilar, metallic or non-metallic materials into a composite whole possessing excellent mechanical and electrical properties; one or more individual operations may be necessary, but the structure, shape, and relative position of the surfaces to be bonded are immaterial."
K. Meyerhans in "Kunststoffe", December 1951, pages 6 and 7: "Casting resin, application and curing: If the thickness of the ARALDITE layer is to be fairly large, Type 1 is no longer suitable on account of its relatively high viscosity. In such cases ARALDITE casting resin B is more suitable...not only for the bonding of components but also for encasing them."
See also the advertisement of Furane Plastics Inc., Los Angeles, a formulator, in "Modern Plastics Encyclopedia for 1962", p. 366: "The Epoxy of a thousand commercial uses: The only all-purpose bonding material, provides super-strength bonds – to practically any surface....: Electrical (Encapsulating, potting, dip coating of connectors, cable terminations, etc.); Building Trades (Repairing cracks..., impregnating..., grouting..., repairing and fabricating...); Marine (Glass cloth lamination..., lining, filling and sealing..., bonding..., potting...); Automotive (Smoothing and filling..., general metal repair); Tooling (Casting..., gelcoating..., filling...)."

[84] In an editorial survey ("The Plastiscope") in the May 1964 issue of "Modern Plastics" R. L. van Boskirk, in the section headed "Epoxy resins continue growth pattern", summed up as follows: "And new markets will continue to be developed and won because epoxies have almost unbelievable bonding and electrical properties and are extremely useful in thousands of relatively small applications." In the July issue the argument is taken a step further ("The Plastiscope", section headed "Epoxy makers expand capacity for specialty needs"): "...The only material with which epoxies could be competitive are polyesters; and any price drop short of reaching the polyester level of the low twenties would have little effect on total epoxy sales. This fact is well recognized by all companies in the field; and they all aim *their sales efforts at those areas where the exceptional properties of epoxy resins make them the only suitable choice.*" (See also Note [4] above.)

[85] Compare L. J. Broutman and F. J. McGarry, respectively Research Assistant and Associate Professor of Materials, Department of Civil Engineering, Massachusetts Institute of Techoulogy (MIT), Cambridge, Mass., in their article "Glass-resin joint strength studies", "Modern Plastics", September 1962, pp. 161 ff. (research performed under Contract No. AF33(616)-6280, Wright Air Development Division, and executed in the MIT Plastics Research Laboratory, supported by the Manufacturing Chemists' Association Inc.): optical (photoelastic) studies with synthetic resin-bonded glass fibres led the authors to the conclusion that "epoxy resins have higher bond strengths than the polyester resins".

[86] See:
A. Puck, Deutsches Kunststoffinstitut, Darmstadt, in: (a) "Kunststoffe", March 1963, article entitled "The dimensions of load-bearing lightweight RP structures". The author concludes that composite materials consisting of resin and glass fibre may be regarded as load-transmitting packages of glass rods bonded together by an adhesive. The resin used was an epoxy. (b) "Kunststoffe", October 1963, article entitled "Glass fibre/resin in heavy-

[87] Vergleiche A. Matting und H. Haferkamp, Institut A für Werkstoffkunde der Technischen Hochschule Hannover in «Kunststoffe», Dez. 1962, Aufsatz «Zum Alterungsverhalten glasfaserverstärkter Kunststoffe»: Die Autoren bemerken, dass «das Alterungsverhalten glasfaserverstärkter Kunststoffe vom Verbundcharakter der Komponenten abhänge.» Spannungsoptische Methoden werden eingesetzt. Bei dynamischer Dauerbelastung verhält sich das Äthoxylin-(Epoxy-)Harz eindeutig günstiger als ein Polyesterharz. Von Interesse sind auch die Hinweise von A. Matting und Kl. Ulmer (Tagung «Metallkleben» in Essen, 20. 10. 60, publ. 1961 im Verlag W. Girardet, Essen), dass nicht nur die eindeutige Definition der sog. Adhäsion d. h. der Bindung zwischen dem «Kunststoffkleber» und dem Metall grosse Schwierigkeiten bereitet, sondern dass auch die prüftechnische Messung bis heute noch nicht gelungen ist. Sie lässt sich nur indirekt über die «Verbindungsfestigkeit» der Klebverbindung (z. B. der Zugscherfestigkeit der einfach überlappten Klebverbindung) abschätzen.

[88] Vergleiche W. Teepe, Technische Hochschule Karlsruhe, Aufsatz: «Mechanische Beanspruchung der Kunststoffe im Massivbau», «Kunststoffe», Nov. 1962: Für kraftübertragende Fugen von grosser Weite werden giessbare mit Steinmaterial gefüllte Äthoxylin-(Epoxy-)Harze – als Giessharzmörtel bezeichnet – eingesetzt. Mittels sog. Ausziehversuche (Abb. 25) wird die Bindefunktion des Harzes messend erfasst. Auch spannungsoptische Methoden werden verwendet.

[89] Report No. A. D. 278 121, Offices of Technical Services, U.S. Dept. of Commerce, Washington 25, «Research and Development for the Improvement of the Strength and Modulus of Continous Filament Wound Cylindrical Structures Through Application of Quality Control», De Bell and Richardson Inc. (referiert in «Materials in Design Engineering», Mai 1963, S. 16): «Flat sided glass fibers replacing round fibers may result in stronger reinforced plastic structures. Several glass hoops 18 in. in dia tested to destruction in a recent study showed that the failure in each case occurred in the non-glass region of the hoop. Flat filaments would make possible nearly 100%-glass content and yet allow uniform bonding area between filaments.»

[90] Es sei z. B. auf die «Annual Technical and Management Conferences, Reinforced Plastics Division» verwiesen, welche in den USA durch die «Society of the Plastics Industry» veranstaltet werden. Für Europa sei an die Jahrestagungen der «Arbeitsgemeinschaft verstärkte Kunststoffe (AVK)» erinnert. (Vergleiche dazu auch »Kunststoffe», Heft 12, Dez. 1964, wo die Vorträge der 4. Öffentlichen Jahrestagung der Arbeitsgemeinschaft Verstärkte Kunststoffe im GKV vom 7.-9. Okt. 1964 in Freudenstadt publiziert sind.)

[91] Vergleiche z. B. K. Ulmer und G. Hennig, Institut A für Werkstoffkunde der Technischen Hochschule Hannover (Dir. Prof. A. Matting). Aufsatz: «Wirtschaftliches Instandsetzen durch Metallkleben», Zeitschrift «Kautschuk und Gummi, Kunststoffe», Nr. 9, 1963, S. 503 ff.: Es werden z. B. mit «Kieselerde gefüllte» Äthoxylin-(Epoxy-)Harze unter Anwendung eines «kombinierten Kleb-Giess-Verfahrens» als «gefüllter Epoxyharz-Klebstoff» zur Abdichtung des Flansches einer Destillationskolonne eingesetzt. Zur messenden Überprüfung der Bindefunktion des Harzes dienen «einschnittig überlappte Zugscherproben und stumpfverklebte Rohrkörper»... Die grundlegende Bedeutung der «einfach überlappten Klebverbindung» kommt z. B. bei den «Verschiebungsmessungen an Metallklebverbindungen», welche A. Matting und U. Krüger (Technische Hochschule Hannover) durchführten, erneut zur Geltung («Kunststoffe», Juni 1964, S. 350ff.)

[92] Vergleiche G. Niederstadt (DFL) im Aufsatz «Neues Prüfverfahren zur Untersuchung der Haftung zwischen Glasfaser und Harz in Glasfaser/Kunststoff-Laminaten» in «Kunststoffe», Juni 1963, S. 355 ff.
Siehe z. B. auch J. E. Carey (Shell Chem. Corp.) am 8th Annual Meeting of the Reinforced Plastics Division, SPI Inc., Febr. 1953, ref. in «Mod. Plast.», 1953, S. 132: Kap.: «Properties of Epoxy Laminates»: «...We have also noted that tensile shear strengths are considerably higher. Tensile shear in this case was measured by pulling apart 1 in. wide test laminates which had been sliced slightly more than half way through on opposite sides at a $\frac{1}{2}$-in. displacement to allow a $\frac{1}{2}$-sq. in. shear area. The epoxy resins show about 4000 to 5000 p.s.i. tensile shear with 181–114 glass cloth using this method, whereas most polyester and diallyl phthalate laminates show about 1000 to 2000 p.s.i. shear strengths». ... So werden auch in einem Referat («Material in Design Engineering», Juli 1964. S. 115, Table 1) über neue Bakelite (Union Carbide)-Epoxyharztypen unter den wenigen aufgezählten typischen Eigenschaften («Typical Properties») Messwerte der «Interlaminar Shear Strength» bekanntgemacht.
Im Aufsatz «Fertigung von Kraftstoff-Zusatzbehältern aus Glasfaserkunststoff» («Luftfahrttechnik-Raumfahrttechnik», 10, 1964, Nr. 4, April), wirft H. E. Meckelburg die Frage auf, ob das in der Metallklebetechnik bereits erfolgreich angewandte Ultraschallprüfverfahren als Grundlage für die Prüfung der Verbundverhältnisse in Glasfaserfabrikaten angewendet werden könnte.

duty lightweight structures": an account of experience acquired with composites (epoxy-bonded) in practical applications.
H. Wurtinger, Deutsches Kunststoffinstitut, Darmstadt, in "Kunststoffe", April 1963, article entitled "The construction and testing of a glider wing using the glass fibre-plastic technique".

[87] See A. Matting and H. Haferkamp, of the Institute A (Materials Technology) of the Hanover Technische Hochschule, in "Kunststoffe", December 1962, article entitled "The ageing characteristics of glass fibre-reinforced plastics": the authors remark that the ageing of glass fibre-reinforced plastics is dependent on the manner in which the components are bonded together. Optical stress methods were used: the behaviour of the epoxy resins was found to be markedly better than that of polyesters when subjected to alternating fatigue tests.
The remarks of A. Matting and K. Ulmer (Conference on Metal Bonding, Essen, 20 October 1960, published in 1961 by Verlag W. Girardet, Essen) concerning adhesion are also of interest. They state that considerable difficulty is encountered not only in formulating an unambiguous definition of "adhesion", i.e. of the bonding action exerted between the plastic adhesive and the metal, but also in its precise measurement. It may only be estimated indirectly, e.g. with the aid of the bonding shear strength test using simple overlapping metal strips.

[88] See W. Teepe, of the Karlsruhe Technische Hochschule, in the article "Mechanical stress in plastics in solid structures", "Kunststoffe", November 1962: pourable epoxy resins filled with stone material and designated "casting resin mortar" were used for making load-transmitting joints where large gaps had to be filled. The bonding function of the resin was measured with the aid of so-called "Ausziehversuche" ("drawing-out tests"; see Illus. 25). Optical stress methods were also employed.

[89] Report No. A.D.278,121, Office of Technical Services, U.S. Dept. of Commerce, Washington 25, D.C., entitled "Research and Development for the Improvement of the Strength and Modulus of Continuous Filament Wound Cylindrical Structures through Application of Quality Control", De Bell and Richardson Inc., reported in "Materials in Design Engineering", May 1963, p. 16: "Flat sided glass fibers replacing round fibers may result in stronger reinforced plastic structures. Several glass hoops 18 in. in dia. tested to destruction in a recent study showed that the failure in each case occurred in the non-glass region of the hoop. Flat filaments would make possible nearly 100% glass content and yet allow uniform bonding area between filaments."

[90] For example the "Annual Technical and Management Conferences, Reinforced Plastics Division" organised by the Society of the Plastics Industry in the U.S.A., or, in Europe, the annual conferences of the Arbeitsgemeinschaft Verstärkte Kunststoffe (AVK).

[91] See, for example, K. Ulmer and G. Hennig, of the Institute A (Materials Technology) of the Hanover Technische Hochschule (Director: Professor A. Matting), in their article "Economical repair techniques utilising metal-bonding agents", "Kautschuk und Gummi, Kunststoffe", No. 9, 1963, pp. 503 ff.: mention is made of a "filled epoxy resin adhesive", described as an epoxy resin filled with silica, that was used for sealing the flange of a distillation column by the technique of "combined bonding/casting". The bonding function of the resin was measured with the aid of "single overlap tensile shear test specimens and butt-bonded tubes". The fundamental importance of the "single overlap adhesive joint" is emphasised elsewhere, e.g. by A. Matting and U. Kruger, of the Hanover Technische Hochschule, in their article ("Measurement of displacement in adhesive joints between metal surfaces)", "Kunststoffe", June 1964, pp. 350 ff.

[92] See G. Niederstadt, of the Deutsche Forschungsanstalt für Luft- und Raumfahrt e.V., in the article "New test procedure for investigation of adhesion between glass fibre and resin in glass fibre-synthetic resin laminates", "Kunststoffe", June 1963, pp. 355 ff.
Compare also J. E. Carey (Shell Chemical Corporation), in his paper read to the 8th Annual Meeting of the Reinforced Plastics Division of the SPI, February 1953, reported in "Modern Plastics", 1953, p. 132; in the section devoted to "Properties of epoxy laminates" he says: "...We have also noted that tensile shear strengths are considerably higher. Tensile shear in this case was measured by pulling apart one-inch-wide test laminates which had been sliced slightly more than half way through on opposite sides at a $^1/_2$-in. displacement to allow a $^1/_2$-sq. in. shear area. The epoxy resins show about 4000 to 5000 p.s.i. tensile shear with 181-114 glass cloth using this method, whereas most polyester and diallyl phthalate laminates show about 1000 to 2000 p.s.i. shear strength...."
In his article "Manufacture of auxiliary fuel tanks using glass-reinforced plastics", published in "Luftfahrttechnik-Raumfahrttechnik", Vol. 10, No. 4, April 1964, H. E. Meckelburg raises the question of whether the ultrasonic techniques that have been successfully adopted for testing the bonding of metals might not form a valid basis for testing the forces of adhesion acting with in glass fibre laminates.

Birkhäuser Verlag Basel und Stuttgart

‹technica›

Internationale technische Zeitschrift
International Technical Review
Revue technique Internationale
Redaktion: H. Buchmann, Basel
21 × 30 cm. Erscheint seit 1952. 26 Nummern jährlich. Abonnementspreis Schweiz Fr. 16.40, Deutschland DM18.40, übriges Ausland Fr. 22.–
Die «technica» orientiert über alle Gebiete der Technik, unter spezieller Betonung der praktischen Belange des allgemeinen Maschinenbaues sowie des Elektromaschinenbaues und der Werkzeugmaschinenindustrie.

‹technica›-Reihe

Hefte1–6
Maschinenkonstruktionslehre
In 5–6 Heften von Prof. Albert Leyer, Professor an der Technischen Hochschule Stuttgart
Heft 1: Allgemeine Gesichtspunkte
48 Seiten mit 87 Abbildungen, Format A4
Fr./DM 7.50 (1963).
Heft 2: Allgemeine Gestaltungslehre
48 Seiten mit 163 Abbildungen, Format A4
Fr./DM 7.50 (1964).
Heft 3: Spezielle Gestaltungslehre, I. Teil
In Vorbereitung
Hefte 4–6: Spezielle Gestaltungslehre, III.–V. Teil

Heft 7
Dieseltriebfahrzeuge der Schweizerischen Bundesbahnen
Von Dipl.-Ing. O. Herrmann, Obermaschineningenieur bei der Kreisdirektion der SBB, Luzern
48 Seiten mit 112 Figuren, Format A4
Fr./DM 6.– (1964)

Heft 8
Gefahrloses Schweissen und Löten von Gas- und Brennstoffbehältern
Von E. Frei-Ischer, Bern
16 Seiten mit 48 Figuren, Format A4
Fr./DM 3.50 (1964)

Technische Werke

Lehrbuch der Schweisstechnik
In zwei Bänden
Von Dr. C. G. Keel

Band I
Lichtbogenschweissverfahren
392 Seiten mit 389 Figuren. Fr./DM 64.– (1961)

Die Grundlagen der Anstrichwissenschaft
Von Dr. A. V. Blom
386 Seiten mit 171 Figuren. Fr./DM 46.– (1954)

Einführung in die Theorie geregelter Gleichstromantriebe
Von Hansruedi Bühler
453 Seiten mit 435 Abbildungen und 7 Tabellen. Fr./DM 58.– (1962)

Einführung in die Anwendung moderner Rechenautomaten
Von Hansruedi Bühler
144 Seiten mit 43 Abbildungen und 26 Tabellen. Fr./DM 23.– (1963)

Giorgis rationales MKS-Maßsystem mit Dimensionskohärenz
für Mechanik, Elektromagnetik, Thermik und Atomistik, fundiert auf Kalantaroffs (LTQ Φ)-System. Von Prof. Dr. Eugen Bodea
2., erweiterte Auflage
142 Seiten mit 17 zum Teil zweiseitigen Tabellen
Fr./DM 28.– (1949)

Röntgenographische Chemie
Von Prof. Dr. E. Brandenberger und Prof. Dr. W. Epprecht
2., erweiterte Auflage, 262 Seiten mit 124 Figuren. Fr./DM 32.– (1960).

Aerodynamik der reinen Unterschallströmung
Von F. Dubs
225 Seiten mit 178 Figuren. Fr./DM 22.– (1954)

Werkstoffkunde der Kerntechnik
Von Prof. Dr. W. Epprecht
407 Seiten mit 141 Figuren. Fr./DM 57.50 (1961)

Die Maschine im Leben der Völker
Von Dr. Franz Maria Feldhaus
341 Seiten mit 205 Figuren und 1 Farbtafel
Fr./DM 26.– (1954)

Raum- und Bauakustik – Lärmabwehr
Von Prof. Willi Furrer
2., erweiterte Auflage. 258 Seiten mit 195 Figuren. Fr./DM 38.50 (1961)

Die Messwandler
Von Dr. phil. und Dipl.-Ing. J. Goldstein
2. Auflage. 222 Seiten mit 220 Figuren
Fr./DM 32.– (1952)

Halbleiterbauelemente
Von Dr. Walter Guggenbühl, Prof. Dr. Ing. Max J.O. Strutt und Dipl.-Ing. Willy Wunderlin

Band I
Halbleiter und Halbleiterdioden
255 Seiten mit 136 Figuren und 12 Tafeln
Fr./DM 38.50 (1962)

Band II
Transistoren
Erscheint 1965

Technische Hydraulik
Von Dr. Charles Jaeger
480 Seiten mit 303 Figuren. Fr./DM 52.– (1949)

Verständliche Chemie
Von Lawrence P. Lessing
Aus dem Englischen übersetzt von Prof. Dr. H. Haas und K. Zingraf
208 Seiten mit 12 Abbildungen
Fr./DM 16.50 (1962)

Probleme der Plastizitätstheorie
Von Prof. Dr. William Prager
100 Seiten mit 52 Figuren. Fr./DM 13.– (1955)

Einführung in die Kontinuumsmechanik
Von Prof. Dr. William Prager
228 Seiten mit 26 Figuren. Fr./DM 32.50 (1961)

Elektrische Maschinen
In fünf Bänden
Von Prof. Dr. R. Richter

Band I
Allgemeine Berechnungselemente
Die Gleichstrommaschinen
Berichtigte und ergänzte Neuauflage
XII und 630 Seiten mit 453 Figuren
Neubearbeitete Auflage erscheint 1965

Band II
Synchronmaschinen und Einankerumformer
3. Auflage. 735 Seiten mit 525 Figuren
Fr./DM 54.– (1963)

Band III
Die Transformatoren
3. Auflage. 349 Seiten mit 252 Figuren
Fr./DM 32.– (1963)

Band IV
Die Induktionsmaschinen
Berichtigte und ergänzte Neuauflage
480 Seiten mit 350 Figuren. Fr./DM 36.40 (1954)

Baustatik für die Praxis
Von Karl Rudmann
128 Seiten mit 123 Figuren. Fr./DM 22.– (1955)

Kleine Technologie des Erdöls
Von Dr. H. Ruf
2., neubearbeitete Auflage
312 Seiten mit 182 Figuren und 56 Tabellen
Fr./DM 42.– (1963)

Elektrizitätslehre
Von PD Dr. Heinz Schilt
216 Seiten mit 187 Figuren. Fr./DM 24.– (1959)

Ins Innere von Kunststoffen, Kunstharzen und Kautschuken
Von Erich V. Schmid
2., erweiterte Auflage, 206 Seiten mit 130 Figuren und 21 Tabellen
Fr./DM 19.– (1949)

Allgemeine Werkstoffkunde
Von PD Dr. Hans Stäger
424 Seiten mit 296 Figuren. Fr./DM 49.– (1947)

Die Gestalt der elektrischen Freileitung
Von Prof. Dr. M. Vidmar
200 Seiten mit 49 Figuren. Fr./DM 20.– (1952)

Die Transformatoren
Von Prof. Dr. M. Vidmar
3., vollständig umgearbeitete Auflage
630 Seiten mit 321 Abbildungen im Text und 2 Tafeln. Fr./DM 68.– (1956)

Elektromagnetische Wellen
Von Prof. Dr. Karl Willy Wagner
267 Seiten mit 185 Figuren und 22 Tabellen
Fr./DM 38.– (1953)

Schaltungstheorie und Messtechnik des Dezimeter- und Zentimeterwellengebietes
Von Dr. Albert Weissfloch
308 Seiten mit 288 Figuren. Fr./DM 34.– (1954)

Einführung in die Hochspannungstechnik
Von Dr. M. Wellauer
328 Seiten mit 281 Figuren. Fr./DM 28.50 (1954)

Mechanik
In drei Bänden
Von Prof. Dr. Hans Ziegler

Band I
Statik der starren und flüssigen Körper sowie Festigkeitslehre
4. Auflage. 244 Seiten mit 389 Figuren
Fr./DM 28.50 (1963)

Band II
Dynamik der starren Körper und Systeme
3. Auflage. 213 Seiten mit 215 Figuren
Fr./DM 29.50 (1962)

Band III
Dynamik der Systeme
2. Auflage. 396 Seiten mit 191 Figuren
Fr./DM 45.– (1956)

If you have any concerns about our products,
you can contact us on
ProductSafety@springernature.com

In case Publisher is established outside the EU,
the EU authorized representative is:
**Springer Nature Customer Service Center GmbH
Europaplatz 3, 69115 Heidelberg, Germany**

Printed by Libri Plureos GmbH
in Hamburg, Germany